Lecture Notes in Mathematics

Edited by A. Dold and B. Eckmann

1107

Nonlinear Analysis and Optimization

Proceedings of the International Conference held in Bologna, Italy, May 3–7, 1982

Edited by C. Vinti

Springer-Verlag
Berlin Heidelberg New York Tokyo 1984

Editor

Calogero Vinti
Dipartimento di Matematica, Università degli Studi
Via Pascoli, 06100 Perugia, Italy

AMS Subject Classification (1980): 49-02, 93-02

ISBN 3-540-13903-6 Springer-Verlag Berlin Heidelberg New York Tokyo
ISBN 0-387-13903-6 Springer-Verlag New York Heidelberg Berlin Tokyo

Printed in Germany

Printing and binding: Beltz Offsetdruck, Hemsbach/Bergstr.
2146/3140-543210

Math Sci
Sep

FOREWORD

In 1980 a very successful Meeting took place at the University of Texas at Arlington (U.S.A.) to celebrate the 70-th birthday of the famous mathematician Lamberto Cesari. Professor Cesari has been for decades in the Faculty of several Institutions of high prestige in the United States, thus honouring the Italian mathematical school, and has always kept close connections with the Italian mathematicians. This led several of his friends and admirers to organize a Meeting in his honour also in Italy and precisely in Bologna, where he was born and was professor at the University for several years before moving to the United States. To this end in the beginning of 1981 an Organizing Committee was formed by Professors D. Graffi (President), V. Boffi (Treasurer), V.E. Bononcini (Secretary), R. Conti, I. Galligani, G. Grioli, R. Nardini (Secretary), A. Pescarini, L. Salvadori, C. Vinti. This Committee obtained cooperation and financial support from several institutions in order to organize an International Meeting in honour of Lamberto Cesari. An Honour Committee was thus formed, composed of (in alphabetical order): Professor Renzo Predi, President, Banca del Monte di Bologna e Ravenna; Professor Carlo Rizzoli, Rettore, Università di Bologna; Professor Raffaele Spongano, President, Accademia delle Scienze dell'Istituto di Bologna; Dott. Lanfranco Turci, President, Regione Emilia-Romagna; Prof. Renato Zangheri, Mayor di Bologna.

Many thanks are due to all members of the Honour Committee for the active interest they took in the organisation of the Meeting, which took place May 3-7, 1982, on the subject "Nonlinear Analysis and Optimization".

The lectures delivered are collected in the present volume.

The Organizing Committee would like to express most sincere thanks also to the Mathematics Committee, C.N.R., in particular to the President Professor Carlo Ciliberto, to the University of Bologna, to the Banca del Monte di Bologna e Ravenna, for the generous financial support which has made the organization of the Meeting and the printing of its Proceedings possible. Thanks are also due to the city of Bologna, to the Regione Emilia-Romagna and to the Accademia delle Scienze dell'Istituto di Bologna where the Meeting actually took place.

Finally, I would like to express my deepest personal thanks to all co-workers of the Organizing Committee and in particular to Professors Boffi, Bononcini, Nardini and Pescarini for their personal commitment.

Dario Graffi

President of the Organizing Committee

TABLE OF CONTENTS

ON THE CONTRIBUTIONS OF LAMBERTO CESARI TO APPLIED MATHEMATICS

D. Graffi
Mathematics Institute, University of Bologna

0. The colleagues of the Organizing Committee have invited me to give the introductory talk to this Meeting and asked to devote it to the scientific activity of Lamberto Cesari. I am very pleased to accept the invitation and I am very grateful to the colleagues for this. I have been friend of Lamberto since 1929, and our friendship has become deeper and deeper as time went by.

Moreover it is not possible in a single talk to describe the whole scientific activity of Cesari; even restricting my consideration to the many areas, both in pure and applied mathematics, to which he brought fundamental contributions, I would end up by a list of results. Therefore I would like to descrive in some detail a couple of questions investigated by him which are well included in the title of the Meeting, namely the nonlinear oscillations and the nonlinear optics or, more generally, wave propagation in nonlinear media.

I will begin by recalling the physical motivation of the above problems: then I will proceed to an overview of the truly remarkable contributions of Cesari. Sometimes, in the course of the exposition, the reformulation of Cesari's result by various people, both in Europe and in U.S.A. (some of them present here) will prove useful: moreover let me state once more that the fundamental ideas belong to him.

Finally, I am sorry if sometimes, for the sake of simplicity, my exposition will necessarily be somewhat imprecise.

1. The nonlinear oscillations theory arise from radiotechnical questions and by now occurs in many branches of electronics. To introduce the description, let us begin by a very elementary example.

Consider an electrical circuit of resistence R, formed by a self induction $\mathcal{L}$, in series with a condensator of capacity C. $\mathcal{L}$,C,R are of course positive. The current $x(t)$ in the circuit satisfies the well known differential equations (which occurs in many other areas of physics).

(1.1) $$\ddot{x}+2p\dot{x}+\omega_o^2 x = 0 \quad , \quad p = R/\mathcal{L}, \quad \omega_o^2 = 1/\mathcal{L}C.$$

It is well known that if $p^2<\omega_o^2$, the solutions of (1.1) are damped oscillations, with damping coefficient p. Radiotelegraphy was first realized by Marconi through oscillations of the above type (or, better through a sequence of such oscillations). However, for radiotelephons and T.V. undamped oscillations are needed or, better, $x(t)$ has to be a periodic function of time. Such a current could be obtained by means of a device governed again by equation (1.1), but with $p=0$.

Experimentally, however, it is difficult to keep a quantity to a constant value. Hence p can vary in time, even though very slowly, around the value zero. Thus for $p>0$ one would get damped oscillations, and amplified oscillations, i.e. oscillations of increasing amplitude, for $p<0$.

It could be imagined to design a device keeping p always negative, even through time-dependent, but this would generate unwanted aperiodic amplified oscillations. It is thus understandable why a device has to be realized such that p is negative for $|x|$ small, so that the oscillations are amplified, but positive for $|x|$ large, so that the oscillations will decrease. Intuitively, some kind of equilibrium condition has to be reached, in which $x(t)$ is periodic.

More precisely, it will be necessary to take p as a function $f(x)$ of x such that $f(x) \lesseqgtr 0$ for $|x| \lessgtr {a \atop b}$ respectively, $0<a\leqslant b$. Then (1.1) becomes

(1.2) $$\ddot{x}+\mu f(x)\dot{x}+\omega_o^2 x = 0$$

where $f(x)$ fulfills the above condition and μ is some positive number to be specified later. (1.2) is of course nonlinear by the occurrence of the term $f(x)$ and is called the Liénard equation; when $f(x)= -(\alpha-\beta x^2)$, $\alpha>0$, $\beta>0$, it becomes the well known Van der Pol equation.

Under very general conditions it can be proved that (1.2) admits periodic solutions; only one (and stable) in the Van der Pol case.
(1.2) is, as already mentioned, an example of nonlinear equation arising from applications. It can be easily generalized in order to include equations often met in further applications: f(x) can be replaced by a function $f(x,\dot{x})$, $\omega_o^2 x$ by a nonlinear function of x, $\varphi(x)$, and terms explicitly depending on t can be added. In the last case the equation is called nonautonomous. For instance if a sinusoidal electromotive force is inserted in the device regulated by (1.2), of amplitude A and frequency ω, one has:

$$\ddot{x}+\mu f(x)\dot{x}+\omega_o^2 x = \mu A \mathrm{sen}\, \omega t \tag{1.3}$$

Furthermore we can consider systems of equations of the type (1.2) and (1.3) and also of more general type.
The need thus arises of a mathematical theory for the class of nonlinear equations just described, which are of practical interest and theoretical as well.

The first investigations on these equations are due to Van der Pol around 1920. He restricted his consideration to the case, of frequent application, of weak non-linearity namely in which the parameter μ is very small with respect to 1. Under this condition working out some approximations, justified also with the help of his remarkable physical intuition, Van der Pol was able to account for the many phenomena occurring in the circuits regulated by (1.2), (1.3).

Van der Pol's investigations have been continued before and after the war by the russian school (Kryloff, Bogoliuboff, Mitropolsky, Minorsky, to whom is due also the credit of making the work of the russian school accessible in the Western world). These authors have implemented general and reliable methods in order to obtain approximate solutions, in the weak nonlinearity case, valid for interval of order $1/\mu$.

Cesari's work comes in around 1960. The investigations of the above authors are the periodic solutions of non linear equations and systems are reconsidered in a completely different way. In the three subsequent sections I will report the main lines of Cesari's investigations; let me recall that those concerning the weak non-linearity case are almost entirely included in his book: "Asymptotic behaviour and stability in ordinary differential equations". Springer Verlag 1963; this book has been translated in Russian.

2. In order to understand Cesari's method, let me start by reconsidering (1.3) written in the following way:

$$\ddot{x}+\omega^2 x = \mu(A\,\mathrm{sen}\,\omega t + \beta x - f(x)\dot{x}), \tag{2.1}$$

$$\beta = \frac{\omega^2 - \omega_o^2}{\mu} \tag{2.2}$$

of course, $\omega^2-\omega_o^2$ is taken of order μ, so that β remains bounded as μ tends to zero .

Perform the Van der Pol transformation on (2.1), i.e. set

(2.3) $x= y_1\mathrm{sen}\,\omega t + y_2\cos\omega t$, (2.4) $\dot{x}=\omega(y_1\cos\omega t-y_2\mathrm{sen}\,\omega t)$

where y_1 and y_2 are unknown functions of t. Equating to second member of (2.4) the derivative of (2.3), a first order equation in $\dot{y}_1$ and $\dot{y}_2$ is obtained. Another equation is obtained inserting $\ddot{x}$, obtained by the derivative of (2.4) in (2.1). Solving the resulting system, one finds:

$$\dot{y}_1 = \frac{\mu}{\omega}(A\,\mathrm{sen}\,\omega t+\beta x-f(x)\dot{x})\cos\omega t\ , \tag{2.5}$$

$$\dot{y}_2 = -\frac{\mu}{\omega}(A\,\mathrm{sen}\,\omega t+\beta x-f(x)\dot{x})\,\mathrm{sen}\,\omega t. \tag{2.6}$$

Inserting in the r.h.s of (2.5) and (2.6), (2.3) and (2.4) one finds a system in y_1 and y_2 which has the advantage of containing the nonlinearity only in the r.h.s..

Let me now consider a vector $\vec{y}(x)$, of components y_1, y_2 and a vector

$\mu\vec{q}(\vec{y},t)$, whose components are the r.h.s. of (2.5) and (2.6) respectively. One has:

$$\dot{\vec{y}} = \mu\vec{q}(\vec{y},t) \tag{2.7}$$

and the vector $\vec{q}(\vec{y},t)$, considered as a function of t, is periodic of period $T = \frac{2\pi}{\omega}$. Remark that $\vec{y}$ and $\vec{q}$ are, in the case examined so far, two-components vectors; however the following considerations are of course valid in the general n-dimensional case.

Thus, by repeated application of the Van der Pol transformations, also systems of second or higher order equations can be reduced to (2.7). It is worth remarking that Cesari has considered also the vector equation obtained adding the vector $B\vec{y}$ to the r.h.s. of (2.7), where B is a matrix. However, we will limit ourselves to (2.7) in what follows.

3. At this point it is necessary to recall some properties of Fourier expansions. Consider the space S of n-components vector functions $\vec{x}(t)=(x_1(t),\dots,x_n(t))$ (we reserve now the symbol $\vec{x}(t)$ to n-dimensional vectors and not to the electric current, unlike Sect. 1), periodic of period T. Let us denote by $\|\vec{x}(t)\|$ the length of the vectors $\vec{x}(t)$, defined in this way.

$$\|\vec{x}(t)\|=|x_1(t)|+|x_2(t)|+\dots+|x_n(t)| \tag{3.1}$$

and introduce in S the norm

$$\upsilon(\vec{x}(t)) = \sup_{t\in(0,T)} \|\vec{x}(t)\|. \tag{3.2}$$

Assume now that $\vec{x}(t)$ can be expanded in Fourier series, and let $\vec{a}_o,\dots,\vec{a}_i,\dots,b_1,\dots,b_i,\dots$, denote its (vector) coefficients, i.e. $(\omega=\frac{2\pi}{T})$

$$\vec{x}(t) = \vec{a}_o + \sum_{1=s}^{\infty} \cos s\omega t\, \vec{a}_s + \text{sen } s\omega t \vec{b}_s \tag{3.3}$$

Now, according to Cesari's notation, set

(3.4) $P_o(\vec{x}(t)) = \vec{a}_o$, (3.5) $P_m(\vec{x}(t)) = \vec{a}_o + \sum_{1}^{m} \cos s\omega t\vec{a}_s + \mathrm{sen}\, s\omega t\vec{b}_s$.

Hence

$$\vec{x}(t) = P_m(\vec{x}(t)) + \sum_{m+1=s}^{\infty} (\cos s\omega t\vec{a}_s + \mathrm{sen}\, s\omega t\vec{b}_s) \; . \tag{3.6}$$

Consider now the expansion:

$$\int (\vec{x}(t) - P_m(\vec{x}(t)))dt = \sum_{m+1=s}^{\infty} \left(\frac{\mathrm{sen}\, s\omega t}{\omega s}\vec{a}_s - \frac{\cos s\omega t}{\omega s}\vec{b}_s\right). \tag{3.7}$$

Remark that the r.h.s. of this equation is nothing else than the antiderivative of the l.h.s. whose average value over a period vanishes; in what follows antiderivatives of this type will be denoted, as in (3.7), by an indefinite integral. Now Cesari proved the following result:

$$\upsilon \int (\vec{x}(t) - P_m(\vec{x}(t)))dt \leqslant C\,\frac{\upsilon(\vec{x}(t))}{\sqrt{m}} \tag{3.8}$$

where C is a positive constant independent of $\vec{x}(t)$.
For m=0, (3.8) takes the form:

$$\upsilon \int (\vec{x}(t) - P_o(\vec{x}(t))dt \leqslant C\,\upsilon(\vec{x}(t)) \tag{3.9}$$

where the value of C in (3.8) and (3.9) is not the same, but it will appear that this does not matter.
It is worth remarking that if:

$$\vec{a}_o = P_o(\vec{x}(t)) \neq 0 \tag{3.10}$$

one has ($\vec{K}$ being some constant)

$$\int \vec{x}(t)dt = \vec{a}_o t + \int (\vec{x}(t) - P_o(x(t)))dt + \vec{K} \; . \tag{3.11}$$

Hence, if $\vec{a}_o \neq 0$ any antiderivative of $\vec{x}(t)$ contains a term linear in t and therefore is not periodic, but its absolute value tends to infinity as $t \to \infty$. Of course, when $\vec{a}_o = P_o(\vec{x}(t)) = 0$, each antiderivative of $\vec{x}(t)$ is periodic.

Let us also remark that

$$(3.12) \qquad \|\vec{a}_o\| < \frac{1}{T} \int_0^T \|\vec{x}(t)\| dt \Rightarrow \|\vec{a}_o\| = \|P_o(\vec{x}(t))\| \leqslant \upsilon(\vec{x}(t)) .$$

4. Consider again (2.7) [where $q(\vec{y}(t),t)$ is a n-component vector function] and let us try to get its periodic solutions through a successive approximation method which is always convergent if μ is small enough, provided the following qualitative conditions are assumed on $\vec{q}(\vec{y},t)$.

For any t and for any $\|\vec{y}(t)\| < R$, ($\|\vec{y}(t)\|$ is defined by (3.1) and R is some positive number) there exist two positive numbers M and L such that, for every t:

$$(4.1) \qquad \|\vec{q}(\vec{y}(t),t)\| \leqslant M , \qquad (4.2) \qquad \|\vec{q}(\vec{y}_1(t),t)-\vec{q}(\vec{y}_2(t),t)\| \leqslant L\|\vec{y}_2(t)-\vec{y}_1(t)\|$$

where, of course, we take $\|\vec{y}_1(t)\| < R$, $\|\vec{y}_2(t)\| < R$.

Let us come back to (2.7). Since μ is assumed very small, it is natural to neglect the term in μ, in first approximation.

Then, one has:

$$(4.3) \qquad \vec{y}_o(t) = \vec{a}$$

where $\vec{a}$ is some constant vector. It is thus natural to take, as second approximation, the quantity:

$$(4.4) \qquad \vec{y}_1^{\,*}(t) = \vec{a} + \mu \int \vec{q}(\vec{y}_o,t)dt .$$

However $\vec{y}_o$ is constant by (4.3) while $\vec{q}(\vec{y}_o,t)$ is a periodic function of time of period T. Now, y_1^* can be periodic only if $P_o(\vec{q}(\vec{y}_o,t)) = 0$: if this is not true, we have already seen that a term linear in t occurs in the r.h.s. of (4.4) (called the secular term according to the accepted terminology). Hence $y_1^*(t)$ cannot be periodic. To overcome this difficulty Cesari, by means of a clever and simple idea (as all clever ideas are), considers a new sequence of vectors $\vec{Z}_k(t)$ (k=0,1,...,n..) without secular terms, defined in the following way:

$$(4.5)\qquad \vec{Z}_o(t)=\vec{a} \quad , \quad \vec{Z}_1(t)=\vec{a} + \mu\int(\vec{q}(\vec{Z}_o(t),t)-P_o(\vec{q}(\vec{Z}_o(t),t)))dt,$$

$$(4.6)\qquad \vec{Z}_k(t)=\vec{a} + \mu\int(\vec{q}(\vec{Z}_{k-1}(t),t)-P_o(\vec{q}(\vec{Z}_{k-1}(t),t)))dt\,.$$

By (3.7) (with m=0) $\vec{Z}_1(t)$ is periodic of period T because the secular term has been eliminated.

Now, remark that if $Z_{k-1}(t)$ is periodic of period T, the same will be true for $\vec{q}(\vec{Z}_{k-1}(t),t)$ and hence, by (4.6) for $Z_k(t)$. Then $Z_k(t)$ is periodic of period T for all k by induction.

Let us now prove that if $|\vec{a}| \leqslant r \leqslant R$, we can take μ in such a way that $\|Z_k\| < R$, for all k.

We proceed again by induction: remark indeed that if $\|Z_{k-1}(t)\| < R$, by (3.9) and (4.1), one has:

$$(4.7)\qquad \mu\upsilon\int(\vec{q}(\vec{Z}_{k-1}(t),t)-P_o(\vec{q}(\vec{Z}_{k-1}(t),t)))dt \leqslant \mu C\upsilon\vec{q}(\vec{Z}_{k-1}(t),t) \leqslant \mu\, C\, M$$

Now, if we take $\mu \leqslant \frac{R-r}{MC}$ the r.h.s. of (4.7) does not exceed R-r. Hence:

$$(4.8)\qquad \|\vec{Z}_K(t)\| \leqslant \|\vec{a}\| + \mu\upsilon\int(\vec{q}(\vec{Z}_{k-1}(t),t)-P_o(\vec{q}(\vec{Z}_{k-1}(t),t)))dt \leqslant$$

$$\leqslant r+R-r = R\,.$$

Therefore, since $\|\vec{Z}_o(t)\|=\|a\| < r$ $\|\vec{Z}_k(t)\| < R$ for all k.

Furthermore (3.9) and (4.2) yield:

$$(4.9)\qquad \upsilon(\vec{Z}_{k+1}(t)-\vec{Z}_k(t)) = \mu\upsilon\int(\vec{q}(\vec{Z}_k(t),t)-\vec{q}(\vec{Z}_{k-1}(t),t) - P_o(\vec{q}(\vec{Z}_k(t),t)-$$

$$-\vec{q}(\vec{Z}_{k-1}(t),t)))dt < \mu C\,\upsilon\,(\vec{q}(\vec{Z}_k(t),t)-\vec{q}(\vec{Z}_{k-1}(t),t)) \leqslant$$

$$\leqslant \mu\, CL\, \upsilon\,(\vec{Z}_k(t)-\vec{Z}_{k-1}(t))\,.$$

Now if μ is chosen not only less than $(R-r)/MC$, but also less than l/CL where $0<l<1$, the sequence of approximations $\{Z_k(t)\}$ fulfills the conditions of the contraction principle and hence $Z_k(t)$ converges to a periodic function $Z(t)$ as $k \to \infty$. Hence, at the limit $k \to \infty$ in (4.6) one has:

$$\vec{Z}(t) = \vec{a} + \int (\vec{q}(\vec{Z}(t),t) - P_o(\vec{q}(\vec{Z}(t),t))) \, dt \tag{4.10}$$

Differentiating this with respect to t, we get that $Z(t)$ is a solution of (2.7) provided that:

$$P_o(\vec{q}(\vec{Z}(t),t)) = 0 \tag{4.11}$$

that is, provided the secular term has been eliminated.
Now $\vec{Z}_k(t)$, and hence $\vec{Z}(t)$, is hence determined by the vector $\vec{a}$.
Hence (4.11) is an equation for the vector $\vec{a}$ (or equivalently, a system for its components $a_1,\dots,a_n$) defined "determining equation" by Cesari.
For $\vec{a}$ fixed and fulfilling (4.11) and μ so small that the conditions of the contraction principle are fulfilled, it can be proved that $\vec{Z}(t)$ is unique. In other words, Cesari has shown that for fixed $\vec{a}$ fulfilling the determining equation there is one and only one periodic solution of (2.7).
It is worth remarking that, since

$$\lim_{k\to\infty} P_o(\vec{q}(\vec{Z}_k(t),t)) = P_o(\vec{q}(\vec{Z}(t),t))$$

the vector $\vec{a}$ and hence $\vec{Z}(t)$ can be determined up to an error of order l^{k+1}, i.e. μ^{k+1}.
As already mentioned, there are further methods to determine periodic solutions of (2.7). Cesari's one, unlike some other methods which require the solution of an equation at each approximation step, has the advantage of reducing the elimination of the secular terms to the solution of a single equation.

Let us remark that there are methods to verify the stability of a periodic solution of (2.7). While we will not discuss the determining equation any further, we limit ourselves to remark that the method allows to investigate some experimentally observed phenomena such as synchronization, frequency demoltiplication and so on.
Let us finally recall that Cesari's method can be applied also to autonomous systems such as the one obtained setting A=0 in (2.1). In this case one has sometimes to assume as unknown β beyond $\vec{a}$, but I cannot give more detail on this subject.

5. Cesari did not limit his investigation to the periodic solutions in the weak nonlinearity case, but has obtained the extension of his methods to more general cases.
Assuming now $\mu=1$ in (2.7) and assuming also the validity of (4.1), (4.2) it is convenient to introduce the following trigonometric polynomial of order m:

$$(5.1) \qquad \vec{x}_o(t) = \vec{a}_o + \sum_{1=s}^{m} (\cos s\omega t\vec{a}_s + \operatorname{sen} s\omega t\vec{b}_s),$$

Generalizing (4.5) and (4.6), Cesari considers the following functions:

$$(5.2) \quad \vec{z}_o(t) = \vec{x}_o(t), \quad (5.3) \quad \vec{z}_k(t) = \vec{x}_o(t) + \int (\vec{q}(\vec{z}_{k-1}(t),t) - P_m(\vec{q}(\vec{z}_{k-1}(t),t))\,dt.$$

Since $P_m(\vec{q}(\vec{z}_k(t),t))$ is nothing else than the sum of the first m terms of the Fourier expansion of $\vec{q}(\vec{z}_k(t),t)$, (3.8) can be applied to the integral in (5.2) and one has:

$$(5.3)_1 \quad \upsilon\int (\vec{q}(\vec{z}_{k-1}(t),t) - P_m(\vec{q}(\vec{z}_{k-1}(t),t)))\,dt \leq \frac{C}{\sqrt{m}}\,\upsilon\,\vec{q}(\vec{z}_{k-1}(t),t)$$

which is analogous to (4.7) with $\frac{1}{\sqrt{m}}$ in place of μ. Taking now m large enough and $\|x_o(t)\| < r$, proceeding as above $\|\vec{z}_k(t)\|$ does not exceed R for all k and thus the contraction principle can be applied to the se-

quence $\{\vec{Z}_k(t)\}$.
Therefore, setting once more $\lim_{k\to\infty} \vec{Z}_k(t)=\vec{Z}(t)$, one has

$$(5.4) \qquad \vec{Z}(t) = \vec{x}_o(t) + \int(\vec{q}(\vec{Z}(t),t) - P_m(\vec{q}(\vec{Z}(t),t)))dt$$

If $\vec{Z}(t)$ has to satisfy (2.7) with $\mu=1$ we must have:

$$(5.5) \qquad \frac{d}{dt}\vec{x}_o(t) = P_m(\vec{q}(Z(t),t)) .$$

Expanding both sides in Fourier series and equating the coefficients of sines and cosines, we get 2m+1 equations in $\vec{a}_o, \vec{a}_1, \ldots, \vec{a}_m$, $\vec{b}_1, \ldots, \vec{b}_m$ which are essentially those of the Galerkin method for solving nonlinear ordinary differential equations: however I cannot insist on that as well as on recent investigations of Cesari and Kannan on nonlinear ordinary differential equations.

6. Let us now turn to nonlinear optics. As already mentioned, Cesari brought essential contributions to this area, which led him to the formulation of a new mathematical problem in partial differential equations Let us begin by describing a typical phenomenon of nonlinear optics. A quartz plate is illuminated by a high intensity, laser emitted light beam, of red colour and wavelength $\lambda=6940\ A^o$: the experiment shows the existence of a violet component in the reflected and transmitted light of wavelength $3470\ A^o$, i.e. exactly $\frac{1}{2}$ of the red wavelength. In other words, if ω is the frequency of the laser light, the light emitted by the plate has frequency 2ω, i.e. the plate works as a frequency duplicator. Let us try to implement a qualitative interpretation of this phenomenon.

Let the plate faces be parallel planes, and let us choose an orthogonal coordinate system Oxyz with origin on a face of the plate, the x-axis oriented along the inward normal to the plate: hence if $a > 0$ is the thickness, the faces have equations $x=0$, $x=a$. Assume the laser located very far from the plate in the $x<0$ half-space so that the light wave emitted by the laser can be assumed as plane and propagating along the x-axis in positive direction.

Assume moreover that the wave is plane in the whole space, and polarized in such a way that the electric field is everywhere directed along the y-axis and the magnetic field along the z-axis, so that we can write :

$$(6.1) \qquad \vec{E} = E(x,t)\vec{j} , \qquad (6.2) \qquad \vec{H} = H(x,t)\vec{k} .$$

In this case the Maxwell equations within the plate, i.e. $x \in (0,a)$ are (assuming zero conductivity)

$$(6.3)_1 \qquad -\frac{\partial H}{\partial x} = \varepsilon_o \frac{\partial E}{\partial t} + \frac{\partial \mathcal{P}}{\partial t} , \qquad (6.3)_2 \qquad -\frac{\partial E}{\partial x} = \mu \frac{\partial H}{\partial t}$$

where ε_o, μ are the vacuum dielectric constant and magnetic permeability, respectively (to simplify the matter we take the plate permeability equal to the vacuum one as it is often the case), $\mathcal{P}$ is the electric polarization.

Assuming the vacuum outside the plate, then (6.3) hold also for $x < 0$, $x > a$ provided $\mathcal{P} = 0$.

Now, in the standard dielectric media $\mathcal{P} = \chi E$, where χ is a constant; considering these equations together with $(6.3)_1$, $(6.3)_2$ a linear system is obtained and consequently it is well known that the plate transmiths only waves of the same wavelength of the laser emitted ones. In order to interpret the frequency duplication one has to write a nonlinear relation between $\mathcal{P}$ and E. The simplest one is of course a parabolic relation: then assuming $\mathcal{P} = 0$ if $E=0$, one has (η stands for some constant)

$$(6.4) \qquad \mathcal{P} = \chi E + \frac{\eta}{2} E^2 .$$

Hence, if we set:

$$(6.5) \qquad \varepsilon = \varepsilon_o + \chi .$$

System $(6.3)_1$, $(6.3)_2$

$$(6.6)_1 \quad -\frac{\partial H}{\partial x} = \varepsilon \frac{\partial E}{\partial t} + \eta E \frac{\partial E}{\partial t}, \qquad (6.6)_2 \quad -\frac{\partial E}{\partial x} = \frac{\partial H}{\partial t}$$

which is a nonlinear system because of the last term in $(6.6)_1$. However (6.6) can still be considered as a linear system if $\eta E \frac{\partial E}{\partial t}$ is interpreted as an additional source of light waves in superposition to the laser emitted ones. Of course this term is unknown, but it can be computed through a perturbation method which however is not matematically rigorous.

It is reasonable to think that the violet light comes out of this term. Since the experiments show that the ratio between the red light and violet light intensities is 10^{12}, it is natural to assume η very small or, more precisely, $\frac{\eta E}{\varepsilon} \ll 1$.

Then if $E_o(x,t)$ is the electric field in the linear case, i.e. for $\eta = 0$, it is natural to assume $E-E_o$ and its time derivative of order η. Therefore, writing:

$$(6.7) \quad \eta E \frac{\partial E}{\partial t} = \eta E_o \frac{\partial E_o}{\partial t} + \eta (E-E_o) \frac{\partial E_o}{\partial t} + \eta \frac{\partial}{\partial t}(E-E_o)E_o + \eta (E-E_o) \frac{\partial}{\partial t}(E-E_o)$$

the last terms of this equation are of order at least η^2 and can thus be neglected with respect to the first and, a fortiori, with respect to $\varepsilon \frac{\partial E_o}{\partial t}$.

We thus get:

$$(6.8) \quad \eta \frac{\partial E}{\partial t} E \simeq \eta \frac{\partial E_o}{\partial t} E_o$$

and since E_o is a sinusoidal function of frequency ω, a simple argument shows that $\eta E_o \frac{\partial E_o}{\partial t}$ is a sinusoidal function of frequency 2ω. The additional source has thus frequency twice the laser one, and this explains the above phenomenon. By similar arguments further phenomena of nonlinear optics can be accounted for.

7. In order to carry out a rigorous investigation of the propagation within the plate, one has there to solve (6.6) or the analogous are obtained for example assuming a more general relation between $\mathcal{P}$ and E. In any case, however, we must provide the correct boundary conditions near x=0, x=a.

To this end, we have to examine the electromagnetic field outside the plate. In the half-space x<0, it is represented by the plane wave emitted by the laser, the incident wave, represented by the electromagnetic field $\vec{E}_i = E_i(x,t)\vec{j}$, $\vec{H}_i = H_i(x,t)\vec{k}$.

We must add to it the field of the reflected wave $E_r(x,t)\vec{j}$, $H_r(x,t)\vec{k}$, propagating along the negative x direction.

Hence, for x<0, one has:

$$(7.1) \qquad \vec{E}(x,t)=(E_i(x,t)+E_r(x,t))\vec{j} \quad , \quad \vec{H}(x,t)=(H_i(x,t)+H_r(x,t))\vec{k} \; .$$

For x>a there are no sources nor reflected waves, but only the transmitted wave propagating along the positive x direction with electromagnetic field $\vec{E}_\tau(x,t)$, $\vec{H}_\tau(x,t)$ such that:

$$(7.2) \qquad \vec{E}_\tau(x,t)=E_\tau(x,t)\vec{j}, \qquad \vec{H}_\tau(x,t)=H_\tau(x,t)\vec{k} \; .$$

Let us once more denote by $\vec{E}(x,t)$, $\vec{H}(x,t)$ the field within the plate given by (6.1), (6.2).

Now on the separation planes x=o, x=a, the Maxwell equations require the continuity of the tangential components of the electric field and of the magnetic field.

Since both are parallel to the yz plane, they coincide with their tangential components. Hence for x=0 and all t, one has:

$$(7.3) \quad E_i(0,t)+E_r(0,t) = E(0,t), \qquad (7.4) \quad H_i(0,t)+H_r(0,t) = H(0,t) \, .$$

As it is known from electromagnetic theory, one has (recall that the two waves travel in opposed directions)

$$(7.5)\qquad H_i(0,t) = \sqrt{\frac{\varepsilon_0}{\mu}}\,E_i(0,t)\ ,\qquad (7.6)\qquad H_r(0,t) = -\sqrt{\frac{\varepsilon_0}{\mu}}\,E_r(0,t).$$

Adding (7.4) to (7.3) multiplied by $\sqrt{\frac{\varepsilon_0}{\mu}}$, by (7.5), (7.6) we have:

$$(7.7)\qquad \sqrt{\frac{\varepsilon_0}{\mu}}\,E(0,t)+H(0,t) = 2\sqrt{\frac{\varepsilon_0}{\mu}}\,E_i(0,t).$$

On the plane $x=a$ one has

$$(7.8)\qquad E(a,t) = E_\tau(a,t)\ ,\qquad (7.9)\qquad H(a,t) = H_\tau(a,t)$$

whence, analogously:

$$(7.10)\qquad \sqrt{\frac{\varepsilon_0}{\mu}}\,E(a,t)-H(a,t) = 0\ .$$

Hence the problem of determining the electromagnetic field within the plate is reduced to solving (6.6) for $E(x,t)$, $H(x,t)$ for $0<x<a$, with the boundary conditions (7.7) and (7.10) in which $E_i(0,t)$ is known. Once the electromagnetic field within the plate is known, by means of the continuity conditions (7.3), (7.4) (7.5), (7.6), (7.8) and (7.9) it is easy to compute the reflected and the transmitted fields.
The problem has been set in this framework essentially by Cesari: in particular, (7.7) and (7.10) are due to him.

8. As shown by Cesari and Bassanini it is convenient to assume as unknowns the expressions reducing to the Riemann invariants when $\eta=0$. I.e. we set

$$(8.1)\qquad E(x,t) = \frac{1}{2}\left(\sqrt{\frac{\mu}{\varepsilon}}\,(Z_1(x,t)+Z_2(x,t)\right)\ ,$$

$$(8.2)\qquad H(x,t) = \frac{1}{2}\,(Z_1(x,t)-Z_2(x,t)).$$

By substituting (8.1) and (8.2) into (6.6), after some calculations here omitted, we have:

$$(8.3)\qquad \left(\frac{\partial Z_1}{\partial x} + \rho_1 \frac{\partial Z_1}{\partial t}\right) + \beta \left(\frac{\partial Z_2}{\partial x} + \rho_1 \frac{\partial Z_2}{\partial t}\right) = 0$$

$$(8.4)\qquad \beta \left(\frac{\partial Z_1}{\partial x} + \rho_2 \frac{\partial Z_1}{\partial t}\right) + \left(\frac{\partial Z_2}{\partial x} + \rho_2 \frac{\partial Z_2}{\partial t}\right) = 0$$

where

$$(8.5)\quad \rho_1 = -\rho_2 = \sqrt{\varepsilon\mu}\,(1+2\Theta)^{\frac{1}{2}}, \qquad (8.6)\quad \beta = \Theta^{-1}\left(\sqrt{1+2\Theta}-1-\Theta\right),$$

$$\Theta = \frac{\eta E(x,t)}{\varepsilon}.$$

The boundary conditions (7.7) and (7.10) become:

$$(8.7)\qquad \frac{1}{2}\left(\sqrt{\frac{\varepsilon_o}{\varepsilon}} + 1\right) Z_1(0,t) + \frac{1}{2}\left(\sqrt{\frac{\varepsilon_o}{\varepsilon}} - 1\right) Z_2(0,t) = 2\sqrt{\frac{\varepsilon_o}{\mu}}\, E_i(0,t),$$

$$(8.8)\qquad \frac{1}{2}\left(\sqrt{\frac{\varepsilon_o}{\varepsilon}} - 1\right) Z_1(a,t) + \frac{1}{2}\left(\sqrt{\frac{\varepsilon_o}{\varepsilon}} + 1\right) Z_2(a,t) = 0.$$

The system (8.3), (8.4), with the boundary conditions (8.7), (8.8) is included in the so called becharacteristic systems (and considered earlier by M. Cibrario Cinquini) which I will now write in the most general form in the unknowns $Z_1, Z_2, \ldots, Z_m$ and variables $x, y_1, y_2, \ldots, y_m$

$$(8.9)\qquad \sum_{1=j}^{m} A_{ij}(x,\vec{y},\vec{z}) \left[\frac{\partial Z_j}{\partial x} + \sum_{1=k}^{r} \rho_{ik}(x,\vec{y},\vec{z}) \frac{\partial Z_j}{\partial y_k}\right] = \varphi_i\,(x,\vec{y},\vec{z})$$

where $i=1\ldots, n$; $\vec{z}$, $\vec{y}$ are the vectors of components $Z_1,\ldots,Z_m$, $y_1,\ldots, y_r$ respectively. (8.9) holds for $x \in (0,a)$ and $\vec{y} \in E^r$. It is further assumed that the determinant $|A_{ij}|$ of A_{ij} is such that $|A_{ij}| > K > 0$. System (8.3), (8.4) is included in (8.9). Set indeed $\varphi_i(x,\vec{y},\vec{z})=0\ \forall i$, $m=2$, $r=1$, $y_1=t$, $A_{11}=A_{22}=1$, $A_{12}=A_{21}=\beta$, so that $|A_{ij}| = 1-\beta^2$, $\rho_{11}=\rho$, $\rho_{21}=-\rho$.

Let us try to write the boundary conditions for the system (8.9), generalizing (8.7), (8.8) which will be linear relations among the Z_i's and known quantities. Let us hence write as boundary conditions the m equations.

$$(8.10) \qquad \sum_{1=j}^{m} C_{ij}(\vec{y})Z_j(a_i,\vec{y})=\psi_i(\vec{y})$$

where $C_{ij}(\vec{y})$, $\psi_i(\vec{y})$ are known functions of $\vec{y}$, a_i are m numbers such that $0\leq a_i\leq a$. In (8.7) and (8.8) m=2, $a_1=0$, $a_2=a$, $y_1=t$, $\psi_1(\vec{y})=2E_i(0,t)\sqrt{\frac{\varepsilon_o}{\mu}}$, $\psi_2(\vec{y})=0$, $C_{11}=C_{12}=C_{21}=C_{22}$ are the (constant) coefficients of (8.7) and (8.8).

It is worth remarking that if each a_i is zero the problem reduces to a Cauchy one.

Excluding this and some further extremely particular cases the problem (8.9), (8.10) is completely new.

Cesari showed a theorem of existence, uniqueness and continuous dependence for this problem.

Let us briefly comment on Cesari's assumptions and results on the system (8.3), (8.4), i.e. on nonlinear optics.

Apart from the boundedness and continuity assumptions which are standard in mathematical physics, let us consider the assumption of the matrix having a dominant main diagonal. It consists in assuming A_{ij} slightly different from δ_{ij}. In our case $A_{ij}=\delta_{ij}$ for i=j $A_{ij}-\delta_{ij}=\beta$, for $i\neq j$, which is near 0 by (8.5) because of the occurrence of the factor η. By the same reason $|A_{ij}|= 1-\beta^2$ is positive, in agreement with the assumption of the system (8.9).

Cesari's theorem holds for not too large value of a, the thickness parameters. Thus can be interpreted by saying that the wave propagating within the plate can turn into a shock wave. However Bassanini proved that, in the cases experimentally observed the plate thickness is such that Cesari's theorem holds. Remark then that if $E_i(0,t)$ i.e. the datum of the problem is periodic of period T, the same will be true for the electromagnetic field within the plate and consequently in the whole space.

Let us conclude by some general remarks.

In the past it was popular to say that existence theorem are of little interest in mathematical physics because the physical problem itself ensures the existence of a solution.

It is easy to reject this assertion: only an existence theorem ensures the possibility of a mathematical implementation of a physical problem. Furthermore a constructive existence theorem can yield an approximated solution of the problem.

Thus Bassanini, starting from Cesari's result, was able to find a convergent method of successive approximation which allows to perform numerical computations justifying the results of Sect. 6.

APPLIED MATHEMATICS AND SCIENTIFIC THOUGHT

J. Serrin

University of Minnesota

The mathematical work of Lamberto Cesari has always had a strong element of concrete reality running throught it, from his earliest work on surface theory to his study of the asymptotic behavior of solutions of ordinary differential equations to his modern research on control theory. In this note I wish to pay tribute to this side of Professor Cesari's career, by setting down several thoughts on the nature of mathematics as it is related to the study of the physical world. At the same time I would particularly offer encouragement to those who see the application of mathematics as a central part of our scientific culture, holding the same possibility of achievement in the future that it has always held in the past.

These remarks will display certain biases, since the time is short and the subject immense. But in addition to noting that personal experience and prejudice cannot be avoided when discussing mathematical applications, let me add one overarching principle: "applied" mathematics should be "good" mathematics, and should be marked by the same clarity which all mathematicians necessarily strive for.

Necessarily, applied mathematics cannot be separated from the physical sciences since the goals of both are essentially similar,but at the same time applied mathematics has a distinguished and special role to play because of its logical imperatives. These work in two principal directions, which are quite distinct from each other. The first involves <u>the reductions of various aspects of the physical world to mathematical formalism</u>. The catch word to describe this direction of inquiry is "modelling", and certainly this terminology is currently fashionable enough to serve as a first approximation.

Nevertheless it carries a certain pedestrian tone, subtly lowering the intellectual level. Indeed, would one describe Newton's laws as mere modelling, or Euler's and Cauchy's discovery of the field theories of deformable media, or the invention of non-Euclidean geometries, or the theory of special relativity? These discoveries rather are paramount applications of mathematical thought, turned onto physical problems. They should serve continually to remind us of the power of disciplined logic in formulating paradigms for the world, and they clearly form a central part of our mathematical heritage. Accordingly I am not wholly in accord with the phrase "model-building", even though I have no simple alternative and will even use it occasionally in what follows. But its casual use, without qualification, can only detract from the position which mathematics should occupy in the physical sciences.

In addition to the development of major concepts, this first direction of inquiry proceeds also through axiomatization of existing theories, the purpose being to clarify and simplify their underlying concepts. One frequently hears that the pursuit of axiomatization is "intellectually sterile" and uninteresting. Whatever may be meant by this, surely it cannot have mathematical content, since mathematics itself is based on coherence, rigor, and care. Indeed, to the extent the mathematical community agrees with critics of axiomatization it weakens its case for centrality in all directions. It is the clarity of thought involved in axiomatization which is the essence; and surely thoroughness and rigour are fundamental not only in pure mathematics but also in the foundations of science itself. Without a clear statement of fundamental principles, it is hardly possible to define alternate positions or to determine where some particular theory requires modifications. Indeed it is precisely the clarity of Newtonian mechanics which presaged both its triumphs and its eventual relativistic modifications.

There is still a third focus of inquiry within the broad content of "modelling", namely the specialization and refinement of existing theories, that is, research directed not at the primary level of raw physical experience but at organizing an already available structure for some new application. Such a process occurs for example when one studies boundary layer theory as a substructure within viscous flow theory, itself a substructure within continuum mechanics. Subsuming all cases, the result of the model-building endeavor is a logically articulated structure of justifiable beliefs about nature.

A second general focus of applied mathematical work, and one no less significant than the first, is the creation and the use of abstract mathematics to treat the analytical problems which arise from the modeling of physical structure. The temptation to label this aspect of applied mathematical thought as "routine problem solving" must be resisted vigorously. Naturally, once this is said it becomes obvious, but it is surprisingly common as an immediate reaction. The antidote is a moment's reflection on the origin of the ideas which are today the heritage of pure mathematics. One might conceivably enumerate and catalogue thos ideas whose genesis lay in some application, but discretion forbids and no doubt space precludes. To begin such a list would be too easy; to end it, too difficult. It is the glory of our subject and the justification for our endeavors. And, I cannot help adding, this aspect of applied mathematics clearly shows how fruitless it must ever be to separate "pure" and "applied" research: even if the extremes are distinct the spectrum includes all shades, as is clearly evident in the work of Lamberto Cesari. Obvious as the point is, I will have to return to it again.

Simply pointing out the contributions to mathematics which have applied mathematical origins is only half the story, however. There is an impressive list of achievement also in the problems which have successfully been treated. From the early developments of celestial mechanics, conformal mapping, and calculus of variations to the more recent focus on control theory, quasi-variational inequalities, singular perturbation problems, regularity of solutions of partial differential equations, scattering theory, Fourier analysis, bifurcation theory, turbulence, the criteria for well-posed problems, and phase transition theory, to form only the most minimal list, the quality and breadth of applied mathematical research becomes evident.

To study a physical problem with the help of mathematics requires critical reflection, an unwillingness to acquiesce in logically imprecise development. Even with purposeful direction, views may change due to accumulated experience, as we criticize, discard, or refine what we have developed in order better to tell the story of reality. Overall, there is simply more to it than guessing that A→B and then proving the conjecture: one must also realize the formulation of A itself, entailing thereby subtle questions of proper concepts and definitions.

In this regard, it is remarkable that the sole discipline of classical physics where serious argument about fundamentals still remains is that of thermodynamics. Even though entropy and energy are now granted almost all pervasive influence, it is nearly impossible to obtain agreement on the two laws of thermodynamics from which these concepts derive. Truesdell considered this a subject which cannot be understood as it is taught, Kac called it a "dehydrated elephant", and Bridgman wrote that thermodynamics is "a branch of physics different from all others, having a palpably verbal feel". Here is a subject essentially untouched by mathematical structure, and the lack of precision shows throughout: after 150 years of development it stands in undeniable contrast to other classical fields of mathematical physics, on whose foundations rested the hands of great mathematicians like Newton, Euler, Cauchy, Hamilton, Maxwell and Hilbert. (The work of Carathéodory in thermodynamics, for all its formal and abstract elegance, does not constitute a complete structure for the subject, nor I think would Carathéodory ever have so claimed). While it is no doubt too soon to claim a realistic part for mathematics in delineating the logical structure of classical thermodynamics - it is conceivable for example that the subject is simply not amenable to the logic which pervades other fields of physics and that the failure to agree is a consequence simply of the complications of material structure - there have nevertheless been recent attempts to formulate the principles of thermodynamic structure in a precise way. Some degree of success has been attained, including at least the understanding of the role of joint systems for the concepts of absolute temperature and the mechanical equivalent of heat, as well as in axiomatizing the notion of reversible and quasi-static processes, and in formulating the energy principles of classical continuum mechanics. These advances not only point to the need for further mathematical study, but show again the important role which applied mathematics can play in the foundations of physics.

I have emphasized the crucial and catalytic nature of applied mathematical thought, but for all this the broad public (and even many scientists) knows little of this role and imagines that the major contribution of mathematics to modern life lies in the ability to perform rapid calculation by computer. For this reason it seems particularly important today that mathematicians make a strong case for the relevance of their subject - in both its pure and applied aspects. I am not thinking only in political terms, though it does appear that eschewing the political arena will court serious losses later. Rather it appears that we should examine the philosophical and empirical roots of our subject as evidence for the centrality which I noted earlier. As a case in point, the very success of mathematical thought in explaining the physical world has prompted several writers, notably E. Wigner and P.C. Hamming to examine the reason for what they consider an unreasonable degree of success. It almost goes without saying, in spite of the subtle arguments advanced, that it will be impossible ever to encompass this metaphysical question. On the other hand, one may argue differently. We need not ask why logic is sufficient for physical understanding, but rather agree that progress of any kind is difficult if not impossible without a logical approach: in other words the questions asked by physics require mathematics as a necessary handmaiden of the advances which are achieved.

Whenever great mathematicians faced the problems of physics they kept the experience of nature before them, though always with the goal of rigorous understanding. Daniel Bernoulli in the eighteenth century stated that "to get profit from physical knowledge it is absolutely necessary to be a mathematician", and Euler has written in his paper on the foundations of fluid mechanics that the results he was uncovering "reveal to us the true laws of nature in all their brilliance". In summary, mathematics can be both necessary and sufficient, bringing order, elegance and beauty to parts of science which otherwise can seem complex, disjoint and confining. And surely it goes without saying that the development of mathematical coherence cannot rely solely on collections of fact, but rather on insight, experience, and the development of appropriate abstraction.

The examples I have noted above, and others equally well-known, should make the point that logical consistency and accuracy are marks equally of pure and applied mathematics, that it is not merely expedient ingenuity which alone holds power. Nevertheless, the impression is widespread that the mathematician, whether in his research or in his teaching, will accept without question a given set of equations describing a physical phenomenon without in the slightest inquiring into their meaning and derivation. On the contrary, he should always clarify for himself, and all the more for his students, the purpose and reality of the model he is studying. In this regard, almost at random I found in Lamberto Cesari's treatise on asymptotic behavior a nice phrase clarifying the physical meaning of linear modelling, and indicative of the sort of things teachers should say to their students: "In a deep sense, all mechanical and physical systems are nonlinear, but they are more and more similar to linear systems the smaller are their displacements from their positions of equilibrium. As soon as amplitudes increase systems become essentially nonlinear". Moreover, he goes on to say, "In a wide range of applications it is in fact only asymptotic stability which is of interest. For plain stability is very labile, and may change into instability by the slightest variation of the physical system, of which the differential problem under discussion is, often, only an approximate representation".

That mathematicians should not be separated from the physical world by artificially erected barriers of "pure" and "applied" might seem a truism. Even so, especially in the last 40-50 years, there has been a tendency to claim particular virtue for "purity" in mathematics, and to hold that research is strongest and best when it is most abstract. This belief has its mirror image in the scientist's view that mathematicians themselves are superfluous, even as he continues to use mathematics as a "tool". Though a scientist or an engineer may personally enjoy pursuing some aspects of mathematical culture, the deductive "proof" method may still seem to him of value mainly for cosmetic purposes, adding little of lasting depth. An interesting passage in a recent paper of William Pohl underlines both the problem and the more significant reality. He writes :

> "The physicist C.N. Yang, in a recent lecture, expressed his surprise that the notion of a fibre bundle, which the quantum field theorists had discovered in a primitive form and with which they had dealt with such great difficulty - it is basic to gauge theories, and hence to the new theories of the electromagnetic force and the weak force - had been discovered and developed previously by the mathematicians. How could mathematicians, who have nothing to do with experiments, discover something about the world? This cannot be explained by chance, since such events have happened often in the history of science. Nor can it be explained according to the view of mathematics held by many contemporary philosophers that mathematics is identical with logic and consists of all hypothetical statements, since of the infinity of possible hypothetical statements, mathematicians have chosen those which are later seen to apply to the world. Rather, it can only be explained, it seems to me, on the assumption that mathematics studies the world. It is true that we look at the world at a higher level of abstraction than other natural scientists and we study various possible worlds; but our possible worlds, sooner or later, seem to reflect aspects of the actual world. Often, of course, we look at aspects of the world discovered by other natural sciences, so that the other sciences enrich and aid mathematics, just as mathematics, in turn, provides the basis of the other sciences."

The crucial point is the unity of mathematical thought, which requires not only an appreciation of the power of abstraction but equally an understanding of the relation of mathematical thought and the physical world.

In concluding these brief remarks, I can do no better than a quotation from Hilbert's famous list of problems for the twentieth century. "While I insist upon rigor in proofs as a requirement for a perfect solution of a problem, I should like, on the other hand, to oppose the opinion that only the concepts of analysis, or even those of arithmetic alone, are susceptible of a fully rigorous treatment. This opinion .....I consider entirely mistaken. Such a one-sided interpretation of the requirement of rigor would soon lead us to ignore all concepts that derive from geometry, mechanics, and physics, to shut off the flow of new material from the outside world... What an important, vital nerve would be cut, were we to root out geometry and mathematical physics!".

In addition Hilbert pointedly discusses the role of axioms in model-building, in a passage which deserves more attention: "...whenever mathematical ideas come up, whether from the theory of knowledge or in geometry, or from the theories of natural science, the task is set for mathematics to investigate the principles underlying these ideas and establish them upon a simple and complete system of axioms in such a way that in exactness and in application to proof the new ideas shall be no whit inferior to the old arithmetical concepts". These words have had more impact than I think is credited to them by pbysicists.

For our part, if we stay alive to the sources of new concepts, if we continue to be interested in the genuine application of genuine mathematical ideas to significant problems, and if we maintain an environment in which interchange of ideas is possible, then we can expect others to remain interested in what we have to say.

Bibliography

[1] KAC M. , Birkhoff Prize Talk, 1978. Mathematical Intelligencer, I, 1978, pp. 97-98.

[2] BONSALL F.F., "A down-to-earth view of mathematics", American Mathematical Monthly, 89, 1982, pp. 8-14.

[3] BROWDER F., Does pure mathematics have a relation to the sciences? (unpublished lecture), pp. 1-29.

[4] CESARI L., Asymptotic behavior and stability problems in ordinary differential equations. Ergebnisse Math. No. 16, Springer-Verlag, Berlin, 1959.

[5] HAMMING R.W., The unreasonable effectiveness of mathematics, American Mathematical Monthly, 87, 1980, pp. 81-90.

[6] JAFFE A., The importance of mathematics, unpublished,1983,pp.1-47.

[7] MURRAY F.J., "Applied Mathematics, An Intellectual Orientation", Plenum Press, New-York-London, 1978. (cf. also Everett Hafner, Bulletin American Mathematical Society, 3, 1980, pp. 735-739.

[8] POHL W., Applications of global differential geometry, Rheinisch-Westfälische Akademie der Wissenschaften, N. 308, 1982, pp. 7-24.

[9] SERRIN J., Conceptual analysis of the classical second laws of thermodynamics, Archive for Rational Mechanics and Analysis, 70, 1979, pp. 355-371.

[10] TRUESDELL C., Is there a philosophy of science?, Centaurus, 17, 1973, pp. 147-161.

[11] TRUESDELL C., The role of mathematics in science as exemplified by the work of the Bernoullis and Euler, Verhandl. Naturf. Ger. Basel, 91, 1980.

[12] TSILIKIS J.D., Simplicity and elegance in theoretical physics, American Scientist, 47, 1959, pp. 87-96.

[13] WIGNER E.P., The unreasonable effectiveness of mathematics in the natural sciences, Comm. Pure Appl. Math., 13, 1960.

NASH POINT EQUILIBRIA FOR VARIATIONAL INTEGRALS

A. Bensoussan and J. Frehse

University Paris Dauphine and INRIA, and
Institut für Angenwandte Mathematik der
Universität Bonn

Introduction

The theory of regularity for the minima of variational integrals has been extensively developed (see M. Giaquinta [3] for an excellent review). It leads to important results, namely that systems of non linear elliptic equations which arise from the calculus of variations, in expressing the stationarity conditions, have regular solutions, whereas this is false for general non linear elliptic systems.

Nevertheless it is important to notice that this is a regularity theory and not an existence theory. In other words the existence of minimum in spaces of less regular functions is in general available by more standard optimization tools.

The situation is quite different in the case of Nash equilibria for variational integrals. Indeed one cannot obtain existence by applying general existence results (like Nash Theorem), the reason being that the conditions are too restrictive.

The objective of this article is to state new results of existence for Nash equilibria as well as regularity results - the basic idea is that they are obtained by combining regularity methods with classical existence results. Namely classical existence results are used for approximated problems in finite dimensional spaces, whereas regularity techniques are used to derive non standard a priori estimates.

The plan is a follows

1. The problem-statement of the results

1.1. Notation - Assumptions

Let Ω be a smooth bounded domain of R^n. We shall consider the Sobolev spaces $H_o^1(\Omega)$, $W_o^{1,p}(\Omega)$, where

$$W_o^{1p}(\Omega) = \{z \in L^p(\Omega) \ , \ \frac{\partial z}{\partial x_i} \in L^p(\Omega)\} \ , \quad p > 1,$$

$$H_o^1 = W_o^{1,2}$$

and W_o^{1p}, H_o^1 denote the closed subspace of functions whose trace on the boundary is o.

Our objective is to study Nash point equilibria for variational integrals. To avoid too many notations, we shall restrict ourselves to two functionals, although all the results extend to an arbitrary finite number of functionals.

Let us consider two real valued functions $F(x,p,q;P,Q)$, $G(x,p,q;P,Q)$ where the arguments are $x \in R^n$, $p,q \in R^N$ $P,Q \in R^{Nn}$. We do not consider the case where p,P vary in R^{N_1}, $R^{N_1 n}$, and q,Q in $R^{N_2}, R^{N_2 n}$, since we want to avoid to work with too many indices. There is no basic simplification if $N_1 = N_2 = N$.

We assume that

(1.1) F,G are Borel functions, continuous in p,q,P Q

(1.2) $|F(x,p,q;P,Q)| \leqq F_o(1+|p|^2 + |q|^2 + |P|^2 + |Q|^2)$

$|G(x,p,q;P,Q)| \leqq G_o(1+|p|^2 + |q|^2 + |P|^2 + |Q|^2)$

(1.3) F is continuously differentiable in p,P,q and

$|F_{p_i}(x,p,q;P,Q)| \leqq F'_o(1 + |p|^2 + |q|^2 + |P|^2 + |Q|^2)$

$|F_{q_i}(x,p,q;P,Q)| \leqq F'_o(1 + |p|^2 + |q|^2 + |P|^2 + |Q|^2)$

$|F_{P_{ij}}(x,p,q;P,Q)| \leqq F''_o(1 + |p| + |q| + |P| + |Q|)$

(1.4) G is continuously differentiable in q,Q,p and

$$|G_{q_i}(x,p,q;P,Q)| \leqq G_o'(1 + |p|^2 + |q|^2 + |P|^2 + |Q|^2)$$

$$|G_{p_i}(x,p,q;P,Q)| \leqq G_o'(1 + |p|^2 + |q|^2 + |P|^2 + |Q|^2)$$

$$|G_{Q_{ij}}(x,p,q;P,Q)| \leqq G_o''(1 + |p| + |q| + |P| + |Q|)$$

(1.5) There exists $\rho > 0$ such that

$$(F_{P_{ij}}(x,p,q;P,Q) - F_{P_{ij}}(x,p,q;\tilde{P},\tilde{Q}))(P_{ij}-\tilde{P}_{ij})$$

$$+ \rho(G_{Q_{ij}}(x,p,q;P,Q) - G_{Q_{ij}}(x,p,q;\tilde{P},\tilde{Q}))(Q_{ij} - \tilde{Q}_{ij})$$

$$\geqq c_o(|P - \tilde{P}|^2 + |Q - \tilde{Q}|^2) \qquad c_o > 0$$

(1.6) $F(x,p,q;P,Q) + G(x,p,q;P,Q) - F(x,\tilde{p},q;\tilde{P},Q)$

$$- G(x,p,\tilde{q};P,\tilde{Q}) \geqq c_1(|P|^2 + |Q|^2) - c_2(|\tilde{P}|^2 + |\tilde{Q}|^2)$$

$$- c_3(|p| + |q| + |\tilde{p}| + |\tilde{q}| + 1)$$

(1.7) $p,P \to F(x,p,q;P,Q)$ is convex

(1.8) $q,Q \to G(x,p,q;P,Q)$ is convex

Here F_o, G_o, F_o', G_o', F_o'', G_o'' are constants not depending on the independent variables.

The assumptions (1.7), (1.8) are satisfied in the following case,

(1.9) F is twice continuously differentiable in p,P and

$$F_{P_{ij}P_{hk}} \geqq c_o I$$

(which is compatible with (1.5))

$$F_{p_i p_h} \geqq \frac{-F_1 |P|^2 I}{1 + |p|^2},$$

$$\left| \sum_i (F_{p_i P_{hk}} P_{hk})^2 \right|^{1/2} \leqq \frac{F_2 |P|^2}{1 + |p|}$$

with $c_o > F_1 + 2F_2$.

Then (1.7) is satisfied, since

$$F_{p_i p_h} p_i p_h + 2F_{p_i P_{hk}} p_i P_{hk} + F_{P_{ij} P_{hk}} P_{ij} P_{hk} \geqq$$

$$c_o |P|^2 - \frac{F_1 |P|^2 |p|^2}{1 + |p|^2} - \frac{2F_2 |P|^2 |p|}{1 + |p|} > 0 .$$

1.2. Examples

Let us consider the following real valued functions

(1.10) $F = a_{ih}^{ik}(x,p,q) P_{ij} P_{hk} + b_{ih}^{jk}(x,q) P_{ij} Q_{hk} + c_{ih}^{ik}(x,p,q) Q_{ij} Q_{hk}$

$+ f_o(x,q,P) + f_1(x,p,q,Q)$

(1.11) $G = \alpha_{ih}^{jk}(x,p,q) Q_{ij} Q_{hk} + \beta_{ih}^{jk}(x,p) Q_{ij} P_{hk} + \gamma_{ih}^{jk}(x,p,q) P_{ij} P_{hk}$

$+ g_o(x,p,Q) + g_1(x,p,q,P)$.

We assume that

(1.12) a_{ih}^{jk} , b_{ih}^{jk} , c_{ih}^{jk} , α_{ih}^{jk} , β_{ih}^{jk} , γ_{ih}^{jk}

are bounded; a, c, α, γ are symmetric

(1.12) $$|f_o| \leqq \overline{f_o}(1 + |q|^2 + |P|)$$

$$|f_1| \leqq \overline{f_1}(1 + |p| + |q|^2 + |Q|)$$

$$|g_o| \leqq \overline{g_o}(1 + |p|^2 + |Q|)$$

$$|g_1| \leqq \overline{g_1}(1 + |p|^2 + |q| + |P|)$$

(1.13) $$\frac{-\bar{a}I}{1 + |p|} \leqq \frac{\partial a_{ih}^{jk}}{\partial p_\alpha} \leqq \frac{\bar{a}I}{1 + |p|} \quad , \quad -\bar{a}I \leqq \frac{\partial a_{ih}^{jk}}{\partial q_\alpha} \leqq \bar{a}I, \ \forall\alpha=1\ldots N$$

$$-\bar{c}I \leqq \frac{\partial c_{ih}^{jk}}{\partial p_\alpha} \quad , \quad \frac{\partial c_{ih}^{jk}}{\partial q_\alpha} \leqq \bar{c}I \quad , \quad \left|\frac{\partial b_{ih}^{jk}}{\partial q_\alpha} P_{ij} Q_{hk}\right| \leqq \bar{b}|P||Q|$$

and an analogous condition for α, β, γ.

(1.14) $$\left|\frac{\partial f_o}{\partial q_\alpha}\right| \leqq \overline{f'_o}(1 + |q|^2 + |P|) \quad , \quad \left|\frac{\partial f_o}{\partial P_{ij}}\right| \leqq \overline{f_o}$$

$$\left|\frac{\partial f_1}{\partial q_\alpha}\right| \leqq f'_1(1 + |p|^2 + |q|^2 + |Q|)$$

$$\left|\frac{\partial f_1}{\partial p_\alpha}\right| \leqq f'_1$$

and analogous relations for g_o, g_1.

(1.15) $$2a_{ih}^{jk}(x,p,q)(P_{hk}-\tilde{P}_{hk})(P_{ij}-\tilde{P}_{ij}) +$$

$$\left(\frac{\partial f_o}{\partial P_{ij}}(x,q,P) - \frac{\partial f_o}{\partial P_{ij}}(x,q,\tilde{P})\right)(P_{ij}-\tilde{P}_{ij})$$

$$+ b^{jk}_{ih}(x,q)(Q_{hk} - \tilde{Q}_{hk})(P_{ij} - \tilde{P}_{ij})$$

$$+ 2\rho\alpha^{jk}_{ih}(x,p,q)(Q_{hk} - \tilde{Q}_{hk})(Q_{ij} - \tilde{Q}_{ij})$$

$$+ \rho(\frac{\partial g_o}{\partial Q_{ij}}(x,p,Q) - \frac{\partial g_o}{\partial Q_{ij}}(x,q,\tilde{Q}))(Q_{ij} - \tilde{Q}_{ij})$$

$$+ \rho\beta^{jk}_{ih}(x,p)(P_{hk} - \tilde{P}_{hk})(Q_{ij} - \tilde{Q}_{ij}) \geqq c_o(|P - \tilde{P}|^2 + |Q - \tilde{Q}|^2).$$

Assuming for instance that f_o, g_o are convex functions of P,Q respectively, that the matrices a and α are uniformly elliptic, and that the matrix β is small, we can find ρ such that (1.115) holds. This implies (1.15).

(1.16) Left hand side of (1.6) =

$$a^{jk}_{ih}(x,p,q)P_{ij}P_{hk} - a^{jk}_{ih}(x,\tilde{p},q)\tilde{P}_{ij}\tilde{P}_{hk}$$

$$+ b^{jk}_{ih}(x,q)(P_{ij}-\tilde{P}_{ij})Q_{hk} + (c^{jk}_{ih}(x,p,q) - c^{jk}_{ih}(x,\tilde{p},q))Q_{ij}Q_{hk}$$

$$+ \Big| \alpha^{jk}_{ih}(x,p,q)Q_{ij}Q_{hk} - \alpha^{jk}_{ih}(x,p,\tilde{q})\tilde{Q}_{ij}\tilde{Q}_{kh}$$

$$+ \beta^{jk}_{ih}(x,p)(Q_{ij}-\tilde{Q}_{ij})P_{hk} + (\gamma^{jk}_{ih}(x,p,q) - \gamma^{jk}_{ih}(x,p,\tilde{q}))P_{ij}P_{hk}\Big|$$

$$+ f_o(x,q,P) - f_o(x,q,\tilde{P}) + f_1(x,p,q,Q) - f_1(x,\tilde{p},q,Q)$$

$$+ g_o(x,p,Q) - g_o(x,p,\tilde{Q}) + g_1(x,p,q,P) - g_1(x,p,\tilde{q},P) .$$

From (1.14) we obtain that the contributions from f_o , f_1, g_o, g_1 are of linear growth. Since we have already assumed that the matrix β is small, the assumption (1.6) is satisfied if

$$\gamma_{ih}^{jk}(x,p,q) - \gamma_{ih}^{jk}(x,p,\tilde{q}) \quad \text{is small.}$$

To check the convexity assumption (1.7), we assume

$$(1.17) \qquad \frac{\partial^2 a_{ih}^{jk}}{\partial p_\alpha \partial p_\beta} f_{ij} P_{hk} p_\alpha p_\beta \geqq - \frac{F_1 |P|^2 |p|^2}{1 + |p|^2} ,$$

$$\frac{\partial^2 c_{ih}^{jk}}{\partial p_\alpha \partial p_\beta} Q_{ij} Q_{hk} p_\alpha p_\beta \geqq 0,$$

$$\frac{\partial^2 f_1}{\partial p_\alpha \partial p_\beta} \geqq 0 .$$

Note that

$$|F_{p_\alpha P_{hk}} P_{hk}| = 2 \, |\frac{\partial}{\partial p_\alpha} a_{ih}^{jk} P_{ij} P_{hk}| \leqq \frac{2a|p|^2}{1 + |p|} \qquad \forall \alpha$$

hence the last condition (1.9) is satisfied with

$$F_2 = 2\bar{a}\sqrt{N} .$$

Provided that $c_o > F_1 + 2F_2$, (1.17) implies (1.9), hence convexity.

1.3. Definition of the problem

Let us consider 2N functions $v_i(x)$, i = 1..N which belong to H_o^1. We will write u, v for the vector function u_i , v_i . Next consider the gradients $\frac{\partial u_i}{\partial x_j}$, $\frac{\partial v_i}{\partial x_j}$, j = 1..n, which belong to $L^2(\Omega)$. We denote by Du the element of R^{Nn} defined by $\frac{\partial u_i}{\partial x_j}$. (Similarly Dv).

From the growth conditions (1.2) then it is clear that the functions $F(x,u,v;Du,Dv)$ and $G(x,u,v;Du,Dv)$ are in $L^2(\Omega)$. We thus set

$$(1.18)\qquad J(u,v) = \int_\Omega F(x,u,v;Du,Dv)\,dx\ ,$$

$$K(u,v) = \int_\Omega G(x,u,v;Du,Dv)\,dx\ .$$

Our objective is to prove the following

Theorem 1.1 Assume (1.1) .. (1.8).
Then there exists a Nash equilibrium point in $H_o^1 x H_o^1$ for J, K. More over this point $\hat{u}$, $\hat{v}$ belongs to $W^{1,q} \times W^{1,q}$ for some $q > 2$.

Remark 1.1 The result stated in Theorem 1.1 cannot be obtained by applying the known existence theorems on Nash equilibrium points. Furthermore, the existence of weak solutions of the Euler equations for the Nash point problem does not follow from the theory of monotone operators or other known results concerning nonlinear elliptic systems. This is due to the fact that F_p, F_q, G_p, G_q have quadratic growth in P,Q. This is one of the reasons why the Nash point approach is of some interest.

2. Approximation

2.1. Preliminaries

We shall need a basis for $W_o^{1,p}$, $p \geqslant 2$. A convenient way to proceed is to consider $H_o^S(\Omega)$ with $\frac{S-1}{n} \geqq \frac{1}{2} - \frac{1}{p}$.

Then $H_o^S \subset W_o^{1,p}$, dense with continuous injection. Consider the eigen functions w_j

$$(2.1)\qquad ((w_j,z))_{H_o^S} = \lambda_j (w_j\ ,\ z) \quad \forall\ z \in H_o^S\ ,$$

which form an orthonomal basis of L^2 (assuming $|w_i|_{L^2} = 1$).

Then $\frac{w_i}{\sqrt{\lambda_i}}$ is an othonomal basis of H_o^S, $w_i \sqrt{\lambda_i}$ is an orthonomal basis of H^{-s}. The w_i also form a "Hilbert basis" for $W_o^{1,p}$. Since the norm in $W_o^{1,p}$ is strictly convex, the projection onto the subset generated by $|w_j,..,w_m|$ is defined in a unique way. We will write

(2.2) $\quad P_m z = \text{proj. of } z \text{ onto } |w_j,..,w_m| \quad (\text{in } W_o^{1,p})$

and

(2.3) $\quad P_m z \to z \text{ in } W_o^{1,p} \quad \text{as } m \to \infty.$

We shall set

$$(2.4) \qquad J_\varepsilon(u,v) = \sum_{i=1}^{N} \varepsilon \|u_i\|^p_{W_o^{1,p}} + J(u,v),$$

$$(2.5) \qquad K_\varepsilon(u,v) = \sum_{i=1}^{N} \varepsilon \|v_i\|^p_{W_o^{1,p}} + K(u;v)$$

which are defined on $W_o^{1,p} \times W_o^{1,p}$. A preliminary result will be the existence of a Nash point for the functionals J_ε and K_ε, in $W_o^{1,p} x W_o^{1,p}$.

2.2. Approximate penalized problem

We first consider a finite dimensional problem. Consider the restriction J_ε^m, K_ε^m of J_ε, K_ε to $|w_j,..,w_m|$, and look for a Nash equilibrium point for J_ε^m, K_ε^m. We have the following

Lemma 2.1 *There exists a Nash equilibrium point* u_ε^m, v_ε^m *for the functionals* J_ε^m, K_ε^m.

Proof We will omit the index ε. Let us first notice that if a Nash point exists, then by definition

$$J_\varepsilon^m(u^m, v^m) \leqq J_\varepsilon^m(o,v^m),$$

$$K_\varepsilon^m(u^m, v^m) \leqq K_\varepsilon^m(u^m,o).$$

Therefore (we shall write $\|u\|^p$ to simplify notation)

$$\varepsilon \| u^m \|^p + \int F(x,u^m,v^m;Du^m,Dv^m)\,dx$$

$$+ \varepsilon \| v^m \|^p + \int G(x,u^m,v^m;Du^m,Dv^m)\,dx$$

$$\leqq \int F(x,o,v^m;o,Dv^m)\,dx + \int G(x,u^m,o;Du^m,o)\,dx$$

and from (1.6), making use of Poincaré's inequality

(2.6) $$c_1' \int_\Omega (|Du^m|^2 + |Dv^m|^2)\,dx \leqq c_3'.$$

From Poincaré's inequality we deduce

$$\int_\Omega (|u^m|^2 + |v^m|^2)\,dx \leqq c_3''.$$

But since $u_i^m = \sum_{\mu=1}^{m} \xi_i^\mu w_\mu$, and the w_μ form an orthonormal basis of L^2, we obtain

(2.7) $$\sum_{i=1}^{N} \sum_{\mu=1}^{m} (\xi_i^\mu)^2 + \sum_{i=1}^{N} \sum_{\vartheta=1}^{m} (\eta_i^\vartheta)^2 \leqq c_3''.$$

We restrict u^m and v^m to satisfy $|u^m| \leqq M$, $|v^m| \leqq M$, which constitutes a compact subset of $|w_j,..w_m|$, and observe that the restricted mappings J^m, K^m are clearly continuous. In view of (1.7), (1.8) the mappings

$$u^m \to J^m(u^m,v^m),$$

$$v^m \to K^m(u^m,v^m)$$

are convex functions. Thus we can apply the classical Nash Theorem (cf. for instance I. Ekeland [2]) to assert that there exists a Nash equilibrium on $|u^m| \leqq M$, $|v^m| \leqq M$.

Let u_M^m, v_M^m be such a point. Using the a priori estimate (2.7), it is easy to check that, letting $M \to \infty$, a subsequence converges to a Nash equilibrium point for the unrestricted functionals J^m, K^m.

2.3. Nash equilibrium for the penalized functionals

We shall now prove the following

Theorem 2.1. Under the assumptions (1.1),..,(1.8) there exists a Nash equilibrium point in $W_0^{1,p} \times W_0^{1,p}$ for the functionals $J_\varepsilon(u,v)$, $K_\varepsilon(u,v)$ defined by (2.4), (2.5).

Proof. Consider the pair u^m, v^m (we omit the index ε). In Lemma 2.1 we have established the estimate

$$(2.8) \qquad \varepsilon \|u^m\|^p + \varepsilon \|v^m\|^p + c_1' \int_\Omega (|Du^m|^2 + |Dv^m|^2)\,dx \leqq c_3'.$$

Hence

$$(2.9) \qquad u^m,\ v^m \text{ remain in a bounded subset of } W_0^{1,p} \times W_0^{1,p}.$$

Thus we can extract a subsequence such that

$$(2.10) \qquad \begin{aligned} &u^m \to u \text{ in } W_0^{1,p} \text{ weakly, } L^p \text{ strongly, and a.e.,} \\ &v^m \to v \text{ in } W_0^{1,p} \text{ weakly, } L^p \text{ strongly, and a.e..} \end{aligned}$$

Moreover, from the estimate

$$J^m(u^m,v^m) \leqq J^m(P_m u, v^m)$$

we obtain

$$(2.11) \qquad \varepsilon \|u^m\|^p + \int F(x,u^m,v^m,Du^m,Dv^m)\,dx \leqq$$

$$\varepsilon \|P_m u\|^p + \int F(x,P_m u,v^m,D(P_m u),Dv^m)\,dx\,.$$

Consider

$$X_1^m = \int (F(x,u^m,v^m,Du^m,Dv^m) - F(x,P_m u,v^m,Du^m,Dv^m)) dx$$

$$= \int_0^1 \int_\Omega F_{Pi} (x, P_m u + \lambda(u^m - P_m u), v^m, Du^m, Dv^m)$$

$$(u_i^m - P_m u_i) X(A_R^m) d\lambda dx$$

$$+ \int_\Omega (F(x,u^m,v^m,Du^m,Dv^m) - F(x,P_m u,v^m,Du^m,Dv^m)) X(A_R^m) dx$$

where $A_R^m = \{x \mid |u^m| + |v^m| + |P_m u| + |P_m v| \leq R\}$, $\bar{A}_R^m$ its complement,

and $X(A)$ denotes the characteristic function of the set A.
We use (1.2), (1.3) to assert that

$$X_1^m \leq \sqrt{N} \, F_0' \int_\Omega (1+R^2 + |Du^m|^2 + |Dv^m|^2) \, |u^m - P_m u| X(A_R^m) \, dx$$

$$+ F_0 \int_\Omega (2+|u^m|^2 + |P_m u|^2 + 2|v^m|^2 + 2|Du^m|^2 + 2|Dv^m|^2) X(A_R^m) dx.$$

From Hölder's inequality, using the fact that u^m, v^m remain in a bounded subset of $W_0^{1,p}$

(2.12) $$|X_1^m| \leq C \left(\int_\Omega |u^m - P_m u|^{p/(p-2)} X(A_R^m) dx \right)^{p/(p-2)} (1+R^2)$$

$$+ C \left(\int_\Omega X(A_R^m) dx \right)^{p/(p-2)}.$$

But

(2.13) $$\int_\Omega X(A_R^m) dx \leq \frac{1}{R^2} \int_\Omega (|u^m| + |v^m| + |P_m u| + |P_m v|)^2 dx \leq \frac{C}{R^2}.$$

Moreover for any fixed R we have

$$\int_\Omega |u^m - P_m u|^{P/(p-2)} X(A_R^m)\,dx \to 0, \quad \text{as} \quad m \to \infty$$

which, together with 2.13) and (2.12) implies that $X_1^m \to 0$, as $m \to \infty$. A similar reasoning proves that

$$X_2^m = \int (F(x,P_m u,v^m,Du^m,Dv^m) - F(x,P_m u,P_m v,Du^m,Dv^m))\,dx$$

$$\to 0 \text{ as } m \to \infty.$$

We set

$$Y_1^m = \int (G(x,u^m,v^m,Du^m,Dv^m) - G(x,u^m,P_m v,Du^m,Dv^m))\,dx$$

$$Y_2^m = \int (G(x,u^m,P_m v,Du^m,Dv^m) - G(x,P_m u,P_m v,Du^m,Dv^m))\,dx$$

and obtain

$$Y_1^m\,,\ Y_2^m \to 0 \quad \text{as} \quad m \to \infty.$$

Similarly

$$Z_1^m = \int (F(x,P_m u,v^m,D(P_m u),Dv^m) - F(x,P_m u,P_m v,D(P_m u),Dv^m))\,dx$$

$$\to 0 \quad \text{as} \quad m \to \infty$$

$$Z_2^m = \int (G(x,u^m,P_m v,Du^m,D(P_m v)) - G(x,P_m u,P_m v,Du^m,D(P_m v))\,dx$$

$$\to 0 \quad \text{as} \quad m \to \infty.$$

We next consider the expressions

$$\Sigma_1^m = \int (F(x,P_m u,P_m v,Du^m,Dv^m) - F(x,P_m u,P_m v,D(P_m u),Dv^m))\,dx$$

$$= \int_\Omega dx \int_0^1 d\lambda \Big(F_{P_{ij}}(x,P_m u,P_m v,D(P_m u)+\lambda(Du^m-D(P_m u)),Dv^m)$$

$$- F_{P_{ij}}(x,P_m u,P_m v,D(P_m u),D(P_m v))\Big)\left(\frac{\partial u_i^m}{\partial x_j} - \frac{\partial P_m u_i}{\partial x_j}\right)$$

$$+ \int F_{P_{ij}}(x,P_m u,P_m v,D(P_m u),D(P_m v))\left(\frac{\partial u_i^m}{\partial x_j} - \frac{\partial (P_m u_i)}{\partial x_j}\right) dx$$

and

$$\Sigma_2^m = \int (G(x,P_m u,P_m v,Du^m,Dv^m)-G(x,P_m u,P_m v,Du^m,D(P_m v)))dx$$

$$= \int_\Omega dx \int_0^1 d\lambda \Big(G_{Q_{ij}}(x,P_m u,P_m v,Du^m,D(P_m v)+\lambda(Dv^m-D(P_m v))$$

$$- G_{Q_{ij}}(x,P_m u,P_m v,D(P_m u),D(P_m v))\Big)\left(\frac{\partial v_i^m}{\partial x_j} - \frac{\partial P_m v_i}{\partial x_j}\right)$$

$$+ \int G_{Q_{ij}}(x,P_m u,P_m v,D(P_m u),D(P_m v))\left(\frac{\partial v_i^m}{\partial x_j} - \frac{\partial P_m v_i}{\partial x_j}\right) dx.$$

By (1.5)

$$\text{(2.14)} \qquad \Sigma_1^m - \Sigma_2^m \geqq \frac{C_o}{2}\int (|Du^m-D(P_m u)|^2+|Dv^m-D(P_m v)|^2)dx$$

$$+\int F_{P_{ij}}(x,P_m u,P_m v,D(P_m u),D(P_m v))\left(\frac{\partial u_i^m}{\partial x_j} - \frac{\partial (P_m u_i)}{\partial x_j}\right) dx$$

$$+ \int G_{Q_{ij}}(x,P_m u,P_m v,D(P_m u),D(P_m v))\left(\frac{\partial v_i^m}{\partial x_j} - \frac{\partial (P_m v_i)}{\partial x_j}\right) dx.$$

Collecting results we obtain

$$\varepsilon \|u^m\|^p + \varepsilon\|v^m\|^p + X_1^m + X_2^m + (Y_1^m + Y_2^m) + \sum_1^m + \sum_2^m$$

$$\leqq \varepsilon \|P_m u\|^p + \varepsilon\|P_m v\|^p + Z_1^m + Z_1^m$$

and from (2.14)

$$(2.15) \qquad \varepsilon\|u^m\|^p + \varepsilon\|v^m\|^p + \frac{C_o}{2} \int (|Du^m - D(P_m u)|^2 + |Dv^m - D(P_m v)|^2)\, dx$$

$$+ \int F_{P_{ij}}(x, P_m u, P_m v, D(P_m u\ D(P_m v)) \left(\frac{\partial u_i^m}{\partial x_j} - \frac{\partial (P_m u_i)}{\partial x_j} \right) dx$$

$$+ \int G_{Q_{ij}}(x, P_m u, P_m v, D(P_m u), D(P_m v)) \left(\frac{\partial v_i^m}{\partial x_j} - \frac{\partial (P_m v_i)}{\partial x_j} \right) dx$$

$$\leqq \varepsilon \|P_m u\|^p + \rho\varepsilon \|P_m v\|^p + S_m$$

where $S_m \to 0$ as $m \to \infty$.

Next we observe that

$$(2.16) \qquad F_{P_{ij}}(x, P_m u, P_m v, D(P_m u), D(P_m v)) \to F_{P_{ij}}(x, u, v, Du, Dv),$$

in L^2.

Indeed there is pointwise convergence and the sequence remains bounded in L^p, $p > 2$. This guarantees (2.16). Since

$$\frac{\partial u_i^m}{\partial x_j} - \frac{\partial P_m u_i}{\partial x_j} \to 0 \quad \text{in } L^2 \text{ weakly}$$

the corresponding integrals to the left side of (2.15) tend to 0 as $m \to \infty$.

Moreover by convexity

$$\lim \|P_m u\|^p - \|u^m\|^p = \|u\|^p - \underline{\lim}\ \|u^m\|^p \leqq 0$$

and we deduce from (2.15) that

$$(2.17) \qquad u^m \to u \quad , \quad v^m \to v \quad \text{in } H_o^1 .$$

The sequence $F(x,u^m,v^m,Du^m,Dv^m)$ converges pointwise and remains bounded in $L^{p/2}$, hence converges in L^1. This implies that the pair u,v is a Nash point for the functionals J_ε , K_ε . The theorem is proved.

3. Nash point for functionals J and K

3.1. Description of the main idea

Our objective in this section is to prove Theorem 1.1. We will consider the Nash point u^ε, v^ε for functionals J_ε, K_ε obtained in Theorem 2.1, and let ε tend to 0. From the estimate (2.81) it is clear that u^ε, v^ε remain in a bounded subset of $H_o^1 \times H_o^1$. However we do not have the bound in $W_o^{1,p}$ which was instrumental in the convergence arguments of Theorem 2.1. Our method will be to obtain an additional estimate in $W^{1,q}$ for some convenient $q > 2$. We will adapt the techniques of Giaquinta-Giusti [5], developed to study the regularity of the minima of variational integrals. They rely on extensions of Gehring's Lemma [3], see Giaquinta-Modica [6], Giaquinta-Giusti [5], Meyers-Elcrat [7].
The basic tool is the following result whose proof can be found in Giaquinta [4].
Let us denote by $Q_R(x_o)$ the n dimensional cube

$$Q_R(x_o) = \{x \in R^n \mid \ |x_i - x_{oi}| < R \quad , \quad i = 1,..,n\} .$$

We use the usual notation

$$\fint_A f dx \equiv \frac{1}{|A|} \int_A f dx .$$

Proposition 3.1. Let Q be an n-cube, and $g \in L^p_{loc(Q)}$, $g \geq 0$, $p > 1$. Assume that

(3.1) $$\fint_{Q_R(x_o)} g^p dx < b\left(\fint_{Q_{2R}(x_o)} g dx\right)^p$$

$$\forall x_o \in Q, \quad \forall R < \text{Min}\left(R_o, \frac{1}{2}\text{dist}(x_o, \partial Q)\right).$$

Then

$$g \in L^q_{loc(Q)}, \quad \text{for } q \in [p, b+\sigma), \text{ and}$$

(3.2) $$\left(\fint_{Q_R} g^q dx\right)^{1/q} \leq c \left(\fint_{Q_{2R}} g^p dx\right)^{1/p}$$

$\forall R$ as above.

The constants σ and c depend only on p, b, n.

An inequality of the type (3.1) is called a reverse Hölder inequality.

3.2. Reverse Hölder inequality

We shall prove the following basic result

Lemma 3.1. The Nash point u^ε, v^ε remains in a bounded set of $(W^{1,q}_o)^2$ for some $q > 2$.

Proof. Let $x_o \in R^n$, consider the Ball $B_R(x_o)$ of radius R and define a vector $\xi^\varepsilon_R = \xi_{R,i}$ $i = 1,..,N$ as follows

$$\xi^\varepsilon_R = 0 \quad \text{if} \quad B_R \cap \partial\Omega \neq \emptyset,$$

$$\xi^\varepsilon_R = u^\varepsilon_R \quad \text{if} \quad B_R \subset \Omega$$

where

$$u_R^\varepsilon = \int_{B_R} u^\varepsilon dx .$$

Replacing u^ε by v^ε, we define quantities ζ_R^ε , v_R^ε . For $0 < t < s < R$, let $\tau(x) \in C^\infty$ satisfy $\tau = 0$ outside B_s , $\tau = 1$ on B_t , $0 \leqq \tau \leqq 1$, $|D\tau| \leqq \frac{C}{s-t}$. The functions

$$\tilde{u} = u^\varepsilon - \tau(u^\varepsilon - \zeta_R^\varepsilon),$$

$$\tilde{v} = v^\varepsilon - \tau(u^\varepsilon - \zeta_R^\varepsilon)$$

belong to $(H_o^1)^N \times (H_o^1)^N$ and are admissible as comparison functions. Therefore

$$\varepsilon \|u^\varepsilon\|^p + \int F(x,u^\varepsilon,v^\varepsilon,Du^\varepsilon,Dv^\varepsilon)dx$$

$$+\rho\varepsilon\|v^2\|^p + \rho\int G(x,u^\varepsilon,v^\varepsilon,Du^\varepsilon,Dv^\varepsilon)dx$$

$$\leqq \varepsilon\|\tilde{u}\|^p + \int F(x,\tilde{u},v^\varepsilon,D\tilde{u},Dv^\varepsilon)dx$$

$$+\rho\varepsilon\|\tilde{v}\|^p + \rho\int G(x,u^\varepsilon,\tilde{v},Du^\varepsilon,D\tilde{v})\ dx .$$

Since $\tilde{u} = u^\varepsilon$, $\tilde{v} = v^\varepsilon$ outside B_s, it remains

$$\varepsilon\int_{B_s} |Du^\varepsilon|^p dx + \int_{B_s} F(x,u^\varepsilon,v^\varepsilon,Du^\varepsilon,Dv^\varepsilon)\ dx$$

$$+ \rho\varepsilon\int_{B_s} |Dv^\varepsilon|^p dx + \rho\int_{B_s} G(x,u^\varepsilon,v^\varepsilon,Du^\varepsilon,Dv^\varepsilon)dx$$

$$\leqq \varepsilon\int_{B_s} |D\tilde{u}|^p dx + \int_{B_s} F(x,\tilde{u},v^\varepsilon,D\tilde{u},Dv^\varepsilon)dx$$

$$+ \rho\varepsilon\int_{B_s} |D\tilde{v}|^p dx + \rho\int_{B_s} G(x,u^\varepsilon,\tilde{v},Du^\varepsilon,D\tilde{v})dx.$$

Using (1.6) we deduce

$$(3.3) \quad \varepsilon\int_{B_s} |Du^\varepsilon|^p dx + \rho\varepsilon\int_{B_s} |Dv^\varepsilon|^p dx + c_1' \int_{B_s} |Du^\varepsilon|^2 dx + \rho c_1' \int_{B_s} |Dv^\varepsilon|^2 dx$$

$$\leqq \varepsilon \int_{B_s} |D\tilde{u}|^p dx + \rho\varepsilon\int_{B_s} |D\tilde{v}|^p dx + c_2' \int_{B_s} |D\tilde{u}|^2 dx + \rho c_2' \int_{B_s} |D\tilde{v}|^2 dx$$

$$+ c_3' |B\ | .$$

But

$$D\tilde{u} = (s-\tau) Du^\varepsilon - D\tau (u^\varepsilon - \xi_R^\varepsilon) .$$

Noting that $s-\tau = 0$ on B_t, we deduce from (3.3)

$$\varepsilon\int_{B_t} |Du^\varepsilon|^p dx + \varepsilon\int_{B_t} |Dv^\varepsilon|^p dx + \int_{B_t} |Du^\varepsilon|^2 dx + \int_{B_t} |Dv^\varepsilon|^2 dx$$

$$\leq K \Big[\varepsilon\int_{B_s - B_t} |Du^\varepsilon|^p dx + \varepsilon\int_{B_s - B_t} |Dv^\varepsilon|^p dx + \int_{B_s - B_t} |Du^\varepsilon|^2 dx$$

$$+ \int_{B_s - B_t} |Dv^\varepsilon|^2 dx$$

$$+ \frac{\varepsilon}{(s-t)^p} \int_{B_R} |u^\varepsilon - \xi_R^\varepsilon|^p dx + \frac{\varepsilon}{(s-t)^p} \int_{B_R} |v^\varepsilon - \zeta_R^\varepsilon|^p dx$$

$$+ \frac{1}{(s-t)^2} \int_{B_R} |u^\varepsilon - \xi_R^\varepsilon|^2 dx + \frac{1}{(s-t)^2} \int_{B_R} |v^\varepsilon - \zeta_R^\varepsilon|^2 dx + |B_R| \Big].$$

Next we fill the hole, which yields

$$(3.4) \quad \varepsilon\int_{B_t} |Du^\varepsilon|^p dx + \varepsilon\int_{B_t} |Dv^\varepsilon|^p dx + \int_{B_t} |Du^\varepsilon|^2 dx + \int_{B_t} |Dv^\varepsilon|^2 dx$$

$$\leq \Theta\Big[\varepsilon\int_{B_t}|Du^\varepsilon|^p dx + \varepsilon\int_{B_s}|Dv^\varepsilon|^p dx + \int_{B_s}|Du^\varepsilon|^2 dx + \int_{B_s}|Dv^\varepsilon|^2 dx\Big]$$

$$+ K_1\Big[\frac{\varepsilon}{(s-t)^p}\int_{B_R}|u^\varepsilon - \xi_R^\varepsilon|^p dx + \frac{\varepsilon}{(s-t)^p}\int_{B_R}|v^\varepsilon - \zeta_R^\varepsilon|^p dx$$

$$+ \frac{1}{(s-t)^2}\int_{B_R}|u^\varepsilon - \xi_R^\varepsilon|^2 dx + \frac{1}{(s-t)^2}\int_{B_R}|v^\varepsilon - \zeta_R^\varepsilon|^2 dx + |B_R|\Big]$$

where $\Theta = \frac{K}{1+K} < 1.$

Using the result stated in Lemma 3.2 below we deduce from (3.4)

$$(3.5)\quad \varepsilon\int_{B_{R/2}}|Du^\varepsilon|^p dx + \varepsilon\int_{B_{R/2}}|Dv^\varepsilon|^p dx + \int_{B_{R/2}}|Du^\varepsilon|^2 dx + \int_{B_{R/2}}|Dv^\varepsilon|^2 dx$$

$$\leq K_2\Big[\frac{1}{R^2}\int_{B_R}|u^\varepsilon - \zeta_R^\varepsilon|^2 dx + \frac{\varepsilon}{R^p}\int_{B_R}|u^\varepsilon - \xi_R^\varepsilon|^p dx$$

$$+ \frac{1}{R^2}\int_{B_R}|v^\varepsilon - \zeta_R^\varepsilon|^2 dx + \frac{\varepsilon}{R^p}\int_{B_R}|v^\varepsilon - \zeta_R^\varepsilon|^p dx + |B_R|\Big].$$

If $B_R \subset \Omega$, $\xi_R^\varepsilon = u_R^\varepsilon$ and the Sobolev Poincaré inequality yields

$$\int_{B_R}|u^\varepsilon - u_R^\varepsilon|^2 dx \leq c\left(\int_{B_R}|Du^\varepsilon|^{\frac{2n}{n+2}}dx\right)^{\frac{n+2}{n}}.$$

If $B_R \cap \partial\Omega \neq \emptyset$, then $\xi_R^\varepsilon = 0$, and

$$\int_{B_R} |u^\varepsilon|^2 dx \leqq \int_{B_{2R}} |u^\varepsilon|^2 dx$$

and by the Sobolev Poincaré inequality

$$\leqq c\left(\int_{B_R} |Du^\varepsilon|^{\frac{2n}{n+2}} dx\right)^{\frac{n+2}{n}}.$$

Therefore in all cases

$$\int_{B_R} |u^\varepsilon - \xi_R^\varepsilon|^2 dx \leqq c\left(\int_{B_{2R}} |Du^\varepsilon|^{\frac{2n}{n+2}} dx\right)^{\frac{n+2}{n}}.$$

Similarly we can assert that

$$\int_{B_R} |u^\varepsilon - \xi_R^\varepsilon|^p dx \leqq c\left(\int_{B_{2R}} |Du^\varepsilon|^{\frac{pn}{n+2}}\right)^{\frac{n+2}{n}} R^{p-2}.$$

Hence we obtain

$$\varepsilon\int_{B_{R/2}} |Du^\varepsilon|^p dx + \varepsilon\int_{B_{R/2}} |Dv^\varepsilon|^p dx + \int_{B_{R/2}} |Du^\varepsilon|^2 dx + \int_{B_{R/2}} |Dv^\varepsilon|^2 dx$$

$$\leqq K_2\left[\frac{1}{R^2}\left(\int_{B_{2R}} |Du^\varepsilon|^{\frac{2n}{n+2}} dx\right)^{\frac{n+2}{n}} + \frac{1}{R^2}\left(\int_{B_{2R}} |Dv^\varepsilon|^{\frac{2n}{n+2}} dx\right)^{\frac{n+2}{n}}\right.$$

$$+ \frac{\varepsilon}{R^2}\left(\int_{B_{2R}} |Du^\varepsilon|^{\frac{pn}{n+2}} dx\right)^{\frac{n+2}{n}} + \frac{\varepsilon}{R^2}\left(\int_{B_{2R}} |Dv^\varepsilon|^{\frac{pn}{n+2}} dx\right)^{\frac{n+2}{n}} + |B_R|\Bigg] .$$

Also this yields

$$(3.6) \qquad \int_{B_{R/2}} (1+|Du^\varepsilon|^2 + |Dv^\varepsilon|^2 + |Du^\varepsilon|^p + |Dv^\varepsilon|^p)\,dx$$

$$\leqq K_3\left(\int_{B_{2R}} (1+ |Du^\varepsilon|^2 + |Dv^\varepsilon|^2 + |Du^\varepsilon|^p + |Dv^\varepsilon|^p)^{\frac{n}{n+2}} dx\right)^{\frac{n+2}{n}},$$

Considering an n dimensional cube $Q \supset \Omega$, and noting that the function

$$g = \left(1 + |Du^\varepsilon|^2 + |Dv^\varepsilon|^2 + \varepsilon|Du^\varepsilon|^p + \varepsilon|Dv^\varepsilon|^p\right)^{\frac{n}{n+2}}$$

belongs to $L_{loc}^{\frac{n+2}{2}}(Q)$ with bounds independant of ε, then by proposition 3.1

$$g \in L^q_{loc}\ (Q)$$

for

$$q \in \left[\frac{n+2}{2}\ ,\ \frac{n+2}{2} + \sigma\right)$$

with bounds independant of ε .

This implies the desired result.

In deducing (3.5) from (3.4) we have used the following

Lemma 3.2. Let $f(t)$ be a non negative bounded function defined in $[\tau_o\ ,\ \tau_1]$, $\tau_o \geqq 0$. Suppose that for $\tau_o \leqq t < s \leqq \tau_1$ we have

$$(3.7)\qquad f(t) \leqq \frac{A_1}{(s-t)^{\alpha}} + \frac{A_2}{(s-t)^{\beta}} + B + \Theta f(s)$$

with constants $0 \leqq \Theta < 1$, A_1, A_2, B, α, β.

Then for all $\tau_o \leqq \rho < R \leqq \tau_1$ we have

$$(3.8)\qquad f(\ ) \leqq c \left[\frac{A_1}{(R-\rho)^{\alpha}} + \frac{A_2}{(R-\rho)^{\beta}} + B \right]$$

with a constant c depending only on α, β, θ.

Proof. Define the sequence

$$t_o = \rho,\ t_{i+1} - t_i = (R-\rho)\tau^i(1-\tau)\ ,\ 0 < \tau < 1.$$

From (3.7) we deduce

$$f(t_i) \leqq \Theta f(t_{i+1}) + \frac{A_1\ \tau^{-i\alpha}}{(R-\rho)^{\alpha}(1-\tau)^{\alpha}} + \frac{A_2\ \tau^{-i\beta}}{(R-\rho)^{\beta}(1-\tau)^{\beta}} + B\ .$$

Multiplying both sides by Θ^i and adding for $i = 0,\ldots,N-1$, we obtain

$$f(\rho) \leqq \Theta^N f(t_N) + B \sum_{i=0}^{N-1} \Theta^i$$

$$+ \frac{A_1}{(R-\rho)^{\alpha}(1-\tau)^{\alpha}} \sum_{i=0}^{N-1} (\Theta\tau^{-\alpha})^i + \frac{A_2}{(R-\rho)^{\beta}(1-\tau)^{\beta}} \sum_{i=0}^{N-1} (\Theta\tau^{-\beta})^i.$$

We choose τ such that $\Theta\tau^{-\alpha} < 1$, $\Theta\tau^{-\beta} < 1$.

Letting N tend to $+\infty$, we deduce (3.8).

Remark 3.1 Lemma 3.2. is an immediate generalization of an easy, but fundamental Lemma used in Giaquinta-Giusti [5].

3.3. Proof of Theorem 1.1

Using the basic estimate of Proposition 3.1., the proof proceeds as in Theorem 2.1: From the sequence u^ε, v^ε, we extract a subsequence converging weakly in $W_o^{1,q} \times W_o^{1,q}$. As in Theorem 2.1 we show that

$$F(x,u^\varepsilon,v^\varepsilon,Du^\varepsilon,Dv^\varepsilon) \to F(x,u,v,Du,Dv)$$

in L^1, and a similar result for G. The theorem follows.

4. Regularity

4.1. General comments

Our objective in this section is to study the regularity properties of the solution u, v which has been obtained in Theorem 1.1. We shall consider a special structure, namely that of the quadratic functionals described in the examples of 1.2..

Our approach will be based on the Euler equations satisfied by u, v. In contrast to the previous sections we have not been able to use the Nash equilibrium conditions themselves to study the regularity. This situation seems to differ from that of minimizing one functional, where it is possible to use the minimizing property directly (cf. M. Giaquinta [4]).

Writing the Euler conditions in the case of the functionals (1.10), (1.11) we arrive at a system of non linear elliptic P.D.E. To distinguish u, v does not make sense any more. That is why we shall consider the following general system

$$(4.1) \qquad -\frac{\partial}{\partial x_i}\left(a_{ij}^{\mu\gamma}(x,u)\frac{\partial u_\gamma}{\partial x_j}\right) = H_\mu(x,u,Du) + f_\mu - \frac{\partial}{\partial x_i} g_i^\mu ,$$

$$u_\mu \in H_o^1(\Omega)$$

where u_μ , $\mu = 1,..,N$ are the unknown functions, and u denotes the vector $(u_1,..,u_N)$; actually u stands for the pair of vectors (u,v) considered previously.

H_μ is a non linear operator involving only the first derivatives of the unknown functions, but which has quadratic growth in Du, and f_μ g_i^μ are given functions in some $L^p(\Omega)$ (in fact in our case $p = +\infty$). We have to be careful in considering the system (4.1), since regularity (say Hölder regularity) does not hold in general for the solution of non linear elliptic systems (cf. M. Struwe |8| for a counter example). We have to take into account that the system (4.1) is related to the Nash equilibrium point of functionals. The main information we shall use is the a priori knowledge that the system (4.1) has a solution in $W_o^{1,q}(\Omega;R^N)$, for some $q > 2$, as a consequence of Gehring's Lemma as previously shown.

4.2. Assumption

We shall assume that

(4.2) $$a_{ij}^{\mu\gamma}(x,u)\xi_\mu^i\xi_\gamma^i \geqq \beta|\xi|^2 , \quad \beta > 0 , \quad \forall \xi ,$$

(4.3) $x,u \to a_{ij}^{\mu\gamma}(x,u)$ are uniformly continuous and bounded in $\Omega x R^N$,

(4.4) $$|H_\mu(x,u,Du)| \leqq K(1 + |Du|^2) ,$$

(4.5) $$g_\mu = 0 \quad , \quad f_\mu = 0 .$$

(4.6) There exists a solution $u \in W_o^{1,q}(\Omega;R^N)$, where q is the exponent arising from Gehring's Lemma.

More precisely, we shall rely on the following reverse Hölder inequality. Let x_o be any point and $B_R(x_o)$ the ball of center x_o and radius R, then we have the property

$$(4.7)\qquad \left(\int_{B_R} |Du|^q\right)^{2/q} \leqq \frac{c\int_{B_{2R}} \left(1 + |Du|^2\right) dx}{R^{n(1-2/q)}}$$

where c is a constant independant of x_o , R.
The property (4.7) is an easy consequence of the estimates obtained in Lemma 3.1.

Remark 4.1. One could relax the assumption of uniform continuity, and also allow a bound in 4.4 depending on u. In fact since the results that we present in this section are known, we have not tried to state the most general theorems. Rather we present a framework which fits the problem we adressed in the previous sections and we describe the most important ideas in order to obtain the regularity, once (4.7) is known. The reader is referred to M. Giaquinta |4| for more details and more general assumptions. Note that by translation we may assume (4.5) without any loss of generality, provided that we allow in (4.4) $K|Du|^2 + h(x)$, with $h \in L^p$, p as large as we wish. This is easily done.

4.3. L^∞ bounds and singular points

In order to prove the regularity we need to know that $u \in L^\infty(\Omega:R^N)$. There are several ways to establish this property. It is possible to assume a structure such that the maximum principle can be applied (see e.g. A. Bensoussan - J. Frehse [1]). However, we shall not dwell on this. Instead we assume a one sided condition. This condition will also guarantee the absence of singular points as we shall see, and this fact is necessary in order to prove the regularity.

Let us assume that in addition to (4.2),..,(4.6)

$$(4.8)\qquad a_{ij}^{\mu\gamma} = a_{ij}\,\delta_{\mu\gamma} ,$$

$$(4.9)\qquad \Sigma_\mu H_\mu u_\mu \leqq \beta_1 |Du|^2 + K \ , \quad \beta_1 < \beta ,$$

where β is the same as in 4.2.

Set

$$A = - \frac{\partial}{\partial x_i} a_{ij} \frac{\partial}{\partial x_j} .$$

In order to define the Green's function relative to the operator A let $Q \Supset \Omega$. For $x_o \in \Omega$, let the Green's function $G = G^{x_o}$ be the solution of

$$(4.10) \qquad \sum_{i,j} \int_Q a_{ij} \frac{\partial z}{\gamma x_j} \frac{\partial G}{\gamma x_i} dx = z(x_o)$$

$$, \forall z \in C_o^{\infty}(Q) ,$$

satisfying $G \in W_o^{1,s}(Q)$, $1 \leqq s < \frac{n}{n-1}$.

We shall need some estimates for G; namely

$$(4.11) \qquad c_o |x-x_o|^{2-n} \leqq G(x) \leqq c_1 |x-x_o|^{2-n}$$

$$\text{(for } n \geqq 3\text{)}.$$

For $n = 2$, one should replace $|x-x_o|^{2-n}$ by $\log |x-x_o|$ in (4.11).
For any $\rho > 0$, one defines G_ρ by solving

$$(4.12) \qquad \sum_{i,j} \int_Q a_{ij} \frac{\partial z}{\partial x_j} \frac{\partial G_\rho}{\partial x_i} dx = \int_{B_\rho} z dx \qquad \forall z \in C_o^{\infty}(Q).$$

G_ρ regularizes G in the following sense

$$(4.13) \qquad G_\rho \to G \text{ in } L^\sigma(Q) , \quad 1 \leqq \sigma < \frac{n}{n-2} ,$$

$$G_\rho \to G \text{ in } W_o^{1,s}(Q) \text{ weakly}, \quad 1 \leqq s < \frac{n}{n-1},$$

$$G_\rho \to G \text{ pointwise } \forall x \neq x_o.$$

We have the following

Lemma 4.1. Assume (4.8), (4.9) (4.9), then one has

$$(4.14)\qquad u \quad L^{\infty}(\Omega;R^N),$$

$$(4.15)\qquad \int_\Omega |Du|^2 G dx \leqq C.$$

Proof. The system (4.1) reduces to

$$(4.16)\qquad -\frac{\partial}{\partial x_i}\left(a_{ij}(x,u)\frac{\partial u_\mu}{\partial x_j}\right) = H_\mu$$

We multiply by $u_\mu G_\rho$ and add, obtaining

$$\sum_\mu \int_\Omega a_{ij} \frac{\partial u_\mu}{\partial x_j} \frac{\partial}{\partial x_i}(u_\mu G_\rho)dx = \sum_\mu \int_\Omega H_\mu u_\mu G_\rho \; .$$

But

$$\int_\Omega a_{ij} \frac{\partial u_\mu}{\partial x_j} u_\mu \frac{\partial G_\mu}{\partial x_i} dx = \frac{1}{2}\int_Q a_{ij} \frac{\partial}{\partial x_j} u_\mu^2 \frac{\partial G_\rho}{\partial x_i} dx$$

$$= \frac{1}{2}\int_{B_\rho} u_\mu^2 dx.$$

Hence we obtain

$$(4.17)\qquad \frac{1}{2}\sum_\mu \int_{B_\rho} u_\mu^2 dx + \sum_\mu \int_\Omega a_{ij} \frac{\partial u_\mu}{\partial x_j}\frac{\partial u_\mu}{\partial x_i} G_\rho dx \leqq \int_\Omega \beta_1 |Du|^2 G_\rho dx + K\int_\Omega G_\rho \, dx$$

hence

$$\sum_\mu \int_{B\rho} u_\mu^2 dx + 2(\beta - \beta_1) \int_\Omega |Du|^2 G_\rho dx \leqq C.$$

Letting ρ tend to 0 and using Fatou's Lemma, we derive (4.15).

Moreover since $\sum_\mu \int_{B_\rho} u_\mu^2 \to \sum_\mu u_\mu^2(x_o)$ a.e., we also deduce (4.14).

Remark 4.2. Suppose we know a priori (4.14) (for instance in a situation where the maximum principle applies), and that the following sign condition holds

$$(4.18) \qquad H_\mu \geqq + c_o \, |Du|^2 - c_1$$

then we easily deduce (4.15) by multiplying (4.16) with G_ρ which yields

$$\int_{B_\rho} u_\mu dx = \int_\Omega U_\mu G_\rho \geqq + \int_\Omega c_o G_\rho |Du|^2 - c_1 \int_\Omega G_\rho .$$

We can also replace (4.18) by

$$H_\mu \leqq - c_o \, |Du|^2 + c_1 .$$

Remark 4.3. Condition (4.15) implies that

$$\int_{B_R} |Du|^2 G dx \to 0$$

as $R \to 0$, hence also

$$(4.19) \qquad \Phi(x_o;R) = R^{2-n} \int_{B_R} |Du|^2 \to 0 \quad \text{as } R \to 0.$$

This expresses the fact that there are no singular points.

4.4. Hölder regularity

We shall follow M. Giaquinta-Giusti [5]. We shall prove Hölder regularity assuming (4.2),..,(4.7) and that (4.14), (4.19) hold for any x_o. From the assumptions it follows that there exists a function $\omega(t,s)$ incrasing in t for fixed s, and in s for fixed t, concave, continuous in (t,o), $\omega(t,o) = 0$ such that

$$(4.20) \qquad |a_{ij}^{\mu\gamma}(x,u) - a_{ij}^{\mu\gamma}(y,v)| \leqq \omega(|x-y|^2 \; , |u-v|^2).$$

We have the following

Lemma 4.2. Take x_o arbitrary, $R \leqq 1$, $\rho < R$, then one has

$$(4.21)\qquad \int_{B_\rho(x_o)} |Du|^2 dx \leqq c_1 \left| \left(\frac{\rho}{R}\right)^n + H(x_o;R) \right| \int_{B_R(x_o)} |Du|^2 dx + c_2 R^n$$

where

$$H(x_o;R) = H\left(R^{2-n} \int_{B_R(x_o)} |Du|^2 dx\right),$$

c_1, c_2 are constants, and H tends to 0 with its argument.

Proof. Since (4.21) is clear if $\rho \geqq \frac{R}{2}$, we may assume $\rho < \frac{R}{2}$. Let u_R denote the mean value of u on the ball B_R. We set

$$a^{\mu\gamma}_{ij,o} = a^{\mu\gamma}_{ij}(x_o, u_R)$$

and consider the function v such that

$$(4.22)\qquad -\frac{\partial}{\partial x_i}\left(a^{\mu\gamma}_{ij,o} \frac{\partial}{\partial x_j} v_\gamma\right) = 0,$$

$$u_\mu - v_\mu \quad H^1_o(B_{R/2}).$$

We have Campanato's estimate

$$(4.23)\qquad \int_B |Dv|^2 dx \leqq c\left(\frac{\rho}{R}\right)^n \int_{B_{R/2}} |Dv|^2 dx .$$

since (4.22) has constant coefficients. Next from (4.22) it follows immediately that

$$(4.24)\qquad \int_{B_{R/2}} |Dv|^2 dx \leqq c \int_{B_{R/2}} |Du|^2 dx .$$

Let w = u - v. From (4.22) and (4.1) we deduce

$$\int_{B_{R/2}} a^{\mu\gamma}_{ij,o} \frac{\partial w_\gamma}{\partial x_j} \frac{\partial w_\mu}{\partial x_i} dx = \int_{B_{R/2}} (a^{\mu\gamma}_{ij}(x_o,u_R) - a^{\mu\gamma}_{ij}(x,u)) \frac{\partial u_\gamma}{\partial x_j} \frac{\partial w_\mu}{\partial x_i} dx$$

$$+ \int_{B_{R/2}} H_\mu x_\mu dx.$$

Hence, using (4.4) and (4.20),

$$(4.25) \quad \int_{B_{R/2}} |Dw|^2 dx \leq c \left| \int_{B_{R/2}} \omega^2 |Du|^2 dx + \int_{B_{R/2}} (1+|Du|^2)|w| dx \right|$$

$$\simeq c \left| R^n + \int_{B_{R/2}} |Du|^2 (|w| + \omega^2) dx \right|$$

by the boundedness of w.

But

$$(4.26) \quad \int_{B_{R/2}} |Du|^2 (|w| + \omega^2) dx < \int_{B_{R/2}} (|Du|^q)^{2/q} dx \int_{B_{R/2}} ((|w| + \omega^2)^{\frac{q}{q-2}})^{\frac{q-2}{q}} dx.$$

From (4.7) and the fact that w, ω are bounded we deduce that the last term can be estimated by

$$\frac{c\int_{B_{2R}} (1+ |Du|^2)\, dx}{R^{n(1-2/q)}} \left(\int_{B_{R/2}} |w| dx + \int_{B_{R/2}} \omega\, dx\right)^{\frac{q-2}{q}}$$

$$= c\int_{B_{2R}} (1+|Du|^2) dx \left(⨍_{B_{R/2}} |w| dx + ⨍_{B_{R/2}} \omega dx\right)^{\frac{q-2}{q}}.$$

But from Poincaré's inequality and Schwarz inequality

$$(4.27)\qquad \fint_{B_{R/2}} |w|\,dx \leqq c\Big(\fint_{B_{R/2}} |Dw|^2 dx\Big)^{1/2}$$

$$\leqq cR\Big(\fint_{B_{R/2}} |Du|^2 dx\Big)^{1/2}.$$

By the concavity of ω, we also have

$$\fint_{B_{R/2}} \omega\, dx \leqq \omega\Big(R^2, \fint_{B_{R/2}} |u-u_R|^2 dx\Big)$$

$$\leqq \omega\Big(R^2, cR^2 \fint_{B_{R/2}} |Du|^2 dx\Big)$$

$$\leqq \omega\Big(1\ , cR^2 \fint_{B_{R/2}} |Du|^2 dx\Big)\ .$$

Collecting results we see that

$$(4.28)\qquad \int_{B_{R/2}} |Du|^2(|w| + \omega^2)dx \leqq c\int_{B_{2R}} (1 + |Du|^2)dx\ H\Big(R^2 \fint_{B_R} |Du|^2 dx\Big)$$

where $H(x) = \sqrt{x} + \omega(1\ , cx)$.

Collecting all the previous estimates we easily deduce (4.21).

We can then assert the following

Theorem 4.1. Assume that (4.2),..,(4.7) and (4.14), (4.19), (4.20) hold. Then the solution u of 4.1 is Hölder continuous.

Proof. Let $\gamma < \alpha < 1$ (arbitrarily close to 1).

We choose $\tau \leq 1$ such that

$$2\, c_1 \tau^{2-2\alpha} \leqq 1.$$

Set

$$c_2\tau^{2-n} = L_o \quad .$$

We can find R_o sufficiently small such that

$$H(\phi(x_o;R)) < \tau^n \qquad \text{for } R < R_o$$

and

$$\phi(x_o;R) + \frac{L_o R^{2\gamma}}{\tau^{2\gamma} - \tau^{2\alpha}} < \varepsilon_o$$

where ε_o is given.

Taking in (4.21) $\rho = \tau R$, $R < R_o$, we obtain

$$\phi(x_o;\tau R) \leqq c_1 \left| 1 + H(x_o;R)\tau^{-n} \right| \tau^2 \phi(x_o;R) + c_2\tau^{2-n}R^{2\gamma}$$

$$\simeq \tau^{2\alpha}\phi(x_o;R) + L_o R^{2\gamma}.$$

Hence

$$\phi(x_o;\tau^k R) \leqq \left| \phi(x_o;R) + \frac{L_o R^{2\gamma}}{\tau^{2\gamma} - \tau^{2\gamma}} \right| \tau^{2k\gamma}$$

$$\leqq \varepsilon_o \tau^{2k\gamma} \quad .$$

Therefore , setting $\rho = \tau^k R$, we obtain

$$\phi(x_o;\rho) \leq \varepsilon_o \left(\frac{\rho}{R}\right)^{2\gamma} \quad \rho < R < R_o.$$

From the classical result of Morrey [8] , it follows that u is in $C^{o,\gamma}(\Omega)$.

Note that γ is arbitrarily close to 1.

Bibliography

[1] BENSOUSSAN A., FREHSE J. - Nonlinear elliptic systems in stochastic game theory. J.Reine Angew. Math. 350 (1984).

[2] EKELAND I. - Théorie des jeux. P.U.F. Paris 1974.

[3] GEHRING F.W. - The L^p-integrability of the partial derivatives of a quasi conformal mapping. Acta Math. 130, 265-277 (1973).

[4] GIAQUINTA M. - Multiple integrals in the calculus of variations and nonlinear elliptic systems. Vorlesungsreihe SFB 72, no. 6, Universität Bonn, 1981.

[5] GIAQUINTA M., GIUSTI E. - On the regularity of the minima of variational integrals. Vorlesungsreihe SFB 72, no. 429, Bonn, 1981.

[6] GIAQUINTA M., MODICA G. - Regularity results for some classes of higher order nonlinear elliptic systems. J. für Reine u. Angew. Math. 311/312, 145-169 (1979).

[7] MEYERS N.G., ELCRAT A. - Some results on regularity for solutions of nonlinear elliptic systems and quasiregular functions. Duke Math. I, 42, 121-136 (1975).

[8] STRUWE M. - A counterexample in elliptic regularity theory. Manuscripta. Math. 34, 85-92 (1981).

NONLINEAR OPTIMIZATION

L. Cesari

Department of Mathematics
University of Michigan - Ann Arbor (Michigan)

We present here existence theorems for multidimensional problems of optimal control and for problems of the calculus of variations concerning
integrals of an extended Lagrangian on a multidimensional domain. We
are particularly interested in the ideas which have led to these theorems. First, problems of optimal control can be deparametrized and reduced to equivalent problems of the calculus of variations with extended Lagragians, and under essentially equivalent general assumptions for
the existence theorems. Second, the underlying lower semicontinuity theorems, or equivalent lower closure theorems, can be proved straightfor-
wardly by an interplay of upper semicontinuity properties of the relevant sets: Kuratowski's property (K), or equivalent lower semicontinuity of the Lagrangian, as sole requirement beside convexity, and conse-
quent stronger property (Q) of suitable auxiliary sets.

1. An extended problem of the calculus of variations

We may consider the problem of minimum of the functional

$$I[x] = \int_G F_o(t,(Mx)(t),(Lx)(t))dt, \quad x \in S,$$

$$F_o(\cdot,(Mx)(\cdot),(Lx)(\cdot)) \in L_1(G), \tag{1}$$

$$(t,(Mx)(t)) \in A, \quad (Lx)(t) \in Q(t,(Mx)(t)), \quad t \in G \ (a.e.),$$

where G is a bounded domain in the R^υ space, $\upsilon \geq 1$, $t = (t^1,...,t^\upsilon) \in G$,
and where the state variable x is thought of as an element of a subset
S of a topological space $(X,\mathcal{C})$, possibly a Sobolev space X on G with its
weak topology $\mathcal{C}$. We think of M: $S \to (L_p(G))^s$, L: $S \to (L_p(G))^r$, $p \geq 1$, as
given operators, not necessarily linear. Thus, we write $y(t)=(Mx)(t) =$

$= (y^1,\dots,y^s)$, $z(t) = (Lx)(t) = (z^1,\dots,z^r)$, $t \in G$.

In (1) A is a given set of the ty-space $\mathbb{R}^{\upsilon+s}$ whose projection on the t-space is G, and for any $(t,y) \in A$ a subset $Q(t,y)$ of the z-space $\mathbb{R}^r$ is given. Thus, in (1) we have constraints on the values taken by the functions $(Mx)(t)$ and $(Lx)(t)$.

If we denote by M_o the set of all $(t,y,z) \in \mathbb{R}^{\upsilon+s+r}$ with $t \in G$, $(t,y) \in A$, $z \in Q(t,y)$, it is convenient to take $F_o = +\infty$ for all $(t,y,z) \in \mathbb{R}^{\upsilon+s+r} - M_o$, thus $F_o(t,y,z) = +\infty$ whenever $t \in \mathbb{R}^{\upsilon} - G$, whenever $t \in G$, $(t,y) \in \mathbb{R}^{\upsilon+s} - A$, and whenever $t \in G$, $(t,y) \in A$, $z \in \mathbb{R}^r - Q(t,y)$. With this convention the constraint $F_o(\cdot,y)(\cdot),z(\cdot)) \in L_1(G)$ implies the remaining constraints. Indeed, $F_o(t,y(t),z(t))$ must be finite for a.a. $t \in G$, hence $(t,y(t)) \in A$, $z(t) \in Q(t,y(t))$ a.e. in G. The same convention $F_o = +\infty$ in $\mathbb{R}^{\upsilon+s+r} - M_o$ will be useful also in the formulation of conditions on F_o.

We denote by $A(t)$ the set of all $y \in \mathbb{R}^s$ such that $(t,y) \in A$.

If we think of S as a subset of a Sobolev space X on G involving derivatives $D^\alpha x$ of orders up to a maximal order N, say, $0 \le |\alpha| \le N$, then Mx could be simply the set of all derivatives $D^\alpha x$ of orders below the maximal order, or $0 \le |\alpha| < N$. Also, S could be the set of all elements $x \in X$ whose boundary values on $\Gamma = \partial G$, or traces $\gamma D^\alpha x$, $0 \le |\alpha| < N$, are given, or satisfy given conditions, say $Bx = 0$.

An existence theorem can now be stated rather easily. To formulate it we state first a general assumption on the extended function F_o (see [1], p. 368): (C^*) $F_o(t,y,z)$ is a given extended function on $\mathbb{R}^{\upsilon+s+r}$ as above, and we assume that, for every $\varepsilon > 0$, there is a compact subset K of G such that (a) meas $(G-K) < \varepsilon$; (b) the extended function F_o restricted to $K \times \mathbb{R}^{s+r}$ is B-measurable; and (c) for almost all $t \in G$ the extended function $F_o(t,y,z)$ of (y,z) has values finite or $+\infty$, and is lower semicontinuous on $\mathbb{R}^{s+r}$.

The function F_o is often called a <u>Lagrangian</u>, and condition (C^*) is just as general as those recently proposed by Ioffe, Ekeland, and Temam. Because of the conventions stated before, condition (C^*) is also a condition on A and the sets $Q(t,x)$.

Under this assumption, the Nemitskii operator appearing in (1), namely $(y,z) \to F_o(t,y(t),z(t))$, maps measurable functions $y(t),z(t)$ into measurable functions $F_o(t,y(t),z(t))$.

We say that an operator $P: S \to (Y,\mathfrak{S})$, $S\subset(X,\mathcal{T})$, has the closure property if $x_k \in S$, $x \in S$, $x_k \to x$ in X, $Px_k \to y$ in Y implies $Px = y$. We say that P has the closed graph property if $x_k \in S$, $x \in X$, $x_k \to x$ in X, $Px_k \to y$ in Y implies $x \in S$, $Px = y$. We say that P has the convergence property provided $x_k \in S$, $x \in X$, $x_k \to x$ in X, implies that there is a subsequence k_s such that $Px_{k_s} \to y$. If $S = X$, closure and closed graph properties coincide. The most usual property that $x_k \in S$, $x \in S$, $x_k \to x$ in X, implies $Px_k \to Px$ in Y is referred to as the continuity of P on S.

(1.i) (A lower semicontinuity theorem). Under condition (C^*), assume that for a.a. $t\in G$ and all $y\in A(t)$, the extended function $F_o(t,y,z)$, $z\in \mathbb{R}^r$, is convex in z. Let $y_k(t), y(t)$, $\xi_k(t)$, $\xi(t)$, $\lambda_k(t)$, $\lambda(t)$, $t\in G$, $k= 1,2,\dots$, be measurable functions on G, with ξ, $\xi_k \in (L_p(G))^r$, $\lambda, \lambda_k \in L_1(G)$, such that $y_k \to y$ in measure in G, $\xi_k \to \xi$ weakly in $(L_1(G))^r$, $\lambda_k \to \lambda$ weakly in $L_1(G)$. Let $\eta_k(t)=F_o(t,y_k(t),\xi_k(t))$, $t \in G$, and assume that $\eta_k(t) \geq \lambda_k(t)$, $t \in G$, $k=1,2,\dots$, and that $-\infty<i = \liminf_k \int_G \eta_k(t)dt<+\infty$. Then there is a function $\eta(t)$, $t\in G$, $\eta \in L_1(G)$, such that $\eta(t) \geq F_o(t,y(t),\eta(t))$ and $\int_G \eta(t)dt \leq i$.

(1.ii) (An existence theorem for the extended problem (1)). Under condition (C^*), assume that for a.a. $t \in G$ and $y \in A(t)$, the extended function $F_o(t,y,z)$, $z \in \mathbb{R}^r$, is convex in z. Let us assume that (λ) for some $\psi(t) \geq 0$, $t \in G$, $\psi \in L_1(G)$, and constant $c \geq 0$ we have $F_o(t,y,z) \geq -\psi(t) - c|z|$. Let S be a nonempty closed set $S = \{x\}$ of elements $x\in X$ with $I[x]$ finite and (c) $S = \{x\}$ is (weakly) relatively compact in X. Assume that both M and L have the closure property, that at least one has the closed graph property, that M has the convergence in measure property, and L has the weak convergence property. Then the functional $I[x]$ in (1) has an absolute minimum in S.

We shall sketch a proof below. A very general Tonelli-type theorem can now be stated as follows:

(1.iii) (A Tonelli-type existence theorem for the extended problem (1)). Same as before, where (λ) is replaced by (ϕ) there is a scalar function $\phi(\zeta)$, $0 \leq \zeta < +\infty$, bounded below, with $\phi(\zeta)/\zeta \to +\infty$ as $\zeta \to +\infty$

such that $F_o(t,y,z) \geq \phi(|z|)$, and (c) is replaced by (c^*) any subclass $S^* = \{x\}^*$ of S with $\{Lx\}^*$ (weakly) relatively compact in $(L_1(G))^s$ is also (weakly) relatively compact in X.

2. A problem of optimal control and associated Lagrangian .

We may consider the problem of minimum with differential equation and constraints

$$\begin{aligned} &I[x,u] = \int_G f_o(t,(Mx)(t),u(t))dt, \; x \in S, \\ &(Lx)(t) = f(t, (Mx)(t), u(t)), \quad t \in G, \qquad (2) \\ &(t,(Mx)(t)) \in A, \; u(t) \in U(t, (Mx)(t)), \; t \in G, \\ &\qquad f_o(\cdot,(Mx)(\cdot),u(\cdot)) \in L_1(G), \end{aligned}$$

where G is a bounded domain in the R^υ space, $\upsilon \geq 1$, $t = (t^1,\dots,t^\upsilon) \in G$, and where the state variable x is thought of as an element of a subset S of a topological space $(X,\mathcal{T})$, possibly a Sobolev space X on G with its weak topology . As before we think of M; $S \to (L_p(G))^s$, L: $S \to (L_p(G))^r$, $p \geq 1$, as given operators, not necessarily linear, and we write $y(t) = (Mx)(t) = (y^1,\dots,y^s)$, $z(t) = (Lx)(t) = (z^1,\dots,z^r)$, $t \in G$. Here, A is a subset of the ty-space $\mathbb{R}^{\upsilon+s}$ whose projection on the t-space is G. For every $(t,y) \in A$ a subset $U(t,y)$ of the u-space $\mathbb{R}^m$ is assigned, $u = (u^1,\dots,u^m)$, and $u(t)$, $t \in G$, above denotes any measurable function on G whose values are on $U(t,y(t))$, or $u(t) \in U(t,y(t))$, $t \in G$. If M denotes the set of all (t,y,u) with $(t,y) \in A$, $u \in U(t,y)$, then f_o scalar and $f = (f_1,\dots,f_r)$ are functions defined on M.

We shall consider the sets

$$Q(t,y) = [z \,|\, z=f(t,y,u),\; u \in U(t,y)] \subset \mathbb{R}^r,$$

$$\tilde{Q}(t,y) = [z^o,z) \,|\, z^o \geq f_o(t,y,u),\; z = f(t,y,u),\; u \in U(t,y)] \subset \mathbb{R}^{r+1},$$

where $(t,y) \in A$, and $Q(t,y)$ is the projection of $\tilde{Q}(t,y)$ on the z-space $\mathbb{R}^r$. Actually, it is convenient to define these sets for every $(t,y) \in \mathbb{R}^{\upsilon+s}$ by taking $Q(t,y) = \emptyset$, $\tilde{Q}(t,y) = \emptyset$, the empty sets, for $(t,y) \in \mathbb{R}^{\upsilon+s} - A$.

We define now the Lagrangian $T(t,y,z)$, $-\infty \leq T(t,y,z) \leq +\infty$, associated to problem (2), by taking

$$T(t,y,z) = \text{Inf}\,[z^o \mid (z^o,z) \in \tilde{Q}(t,y)]$$
$$= \text{Inf}\,[z^o \mid z^o \geq f_o(t,y,u),\ z = f(t,y,u),\ u \in U(t,y)]. \tag{3}$$

where, for $(t,y) \in A$ and $z \in Q(t,y)$ we have $-\infty \leq T < +\infty$, and for $(t,y) \in A$, $z \in \mathbb{R}^r - Q(t,y)$, we have $T = +\infty$, and certainly we have $T = +\infty$ for $(t,y) \in \mathbb{R}^{\upsilon+s} - A$. In other words, T is an extended function defined in $\mathbb{R}^{\upsilon+s+r}$. We shall now associate to problem (2) of optimal control, the new problem of minimum

$$J[x] = \int_G T(t,(Mx)(t),(Lx)(t))dt,\ x \in S,$$
$$T(\cdot,(Mx)(\cdot),(Lx)(\cdot)) \in L_1(G), \tag{4}$$

with the implied constraints

$(t,(Mx)(t)) \in A$, $(Lx)(t) \in Q(t,(Mx)(t))$, $t \in G$ (a.e.).

This is a problem of the calculus of variations of the type (1). We say that we have deparametrized problem (2).

Under mild assumptions, problems (2) and (4) are equivalent, as we shall state below in more details. First, some basic assumptions on the functions f_o, f. A rather general assumption on f_o, f (see [1], p. 385) is the following one:

(C'^*) Given $\varepsilon > o$ there is a compact subset K of G such that (a) meas $(G-K) < \varepsilon$; (b) the sets $A_K = [(t,y) \in A \mid t \in K]$, $M_K = [(t,y,u) \in M \mid t \in K]$ are closed; (c) f_o, f are continuous on M_K.

Under this assumption the Nemitskii operators appearing in (2), namely $(y,u) \to f_o(t,y(t),u(t))$, $(y,u) \to f(t,y(t),u(t))$ map measurable functions y,u into measurable functions. We shall also assume that for almost all $t \in G$ and all (y,z) we have $T(t,y,z) > -\infty$.

We need also state some general properties of the sets $Q(t,y)$. We say that the sets $\tilde{Q}(t,y)$ have the Kuratowski property of upper semicontinuity, or property (K) with respect to y, provided for every (t_o,y_o) we have

$$\tilde{Q}(t_o,y_o) \supseteq \bigcap_{\delta>o} \text{cl} \bigcup_{y\in N_\delta(y_o)} \tilde{Q}(t_o,y). \tag{5}$$

Equivalently, we may say that the set valued map $(t,y) \to \tilde{Q}(t,y)$ has this property. Sets having this property are necessarily closed.
Then, it is well known ([1], p. 294) that the sets $\tilde{Q}(t,y)$ have property (K) with respect to y, if and only if the sets of points (graph) $[(y,z^o,z) \mid (z^o,z)\in \tilde{Q}(t,y),\ y\in A(t)]$ are closed. Also, the sets $\tilde{Q}(t,y)$ have property (K) with respect to y if and only if the extended function $T(t,y,z)$, $(t,y,z) \in \mathbb{R}^{\nu+s+r}$, is lower semicontinuous in (y,z). If this is the case, then all sets $\tilde{Q}(t,y)$ are closed, Inf can be replaced by Min in the definition of T whenever T is finite, and $\tilde{Q}(t,y) = \text{epi}_z\ T(t,y,z)$, that is, $\tilde{Q}(t,y)$ is the epigraph of $T(t,y,z)$, or $\tilde{Q}(t,y) = [(z^o,z) \mid z^o \geq T(t,y,z)]$. If this occurs, and condition (C'^*) holds, then, on the basis of these remarks and of McShane's and Weinberg's implicit function theorem (cf. [1], pp. 275-280), problems of minimum (2) and (4) are equivalent. Moreover, the assumptions we have made guarantee that $T(t,y,z)$ is an extended function satisfying the generic assumtpions we made on F_o in Section 1. In other words, (2) is an extended problem of the calculus of variations as considered in Section 1.

(2.i). (<u>A lower closure theorem</u>) Let G be a bounded region in the t-space $\mathbb{R}^\nu$, $\nu \geq 1$, let A be a subset of the ty-space $\mathbb{R}^{\nu+s}$ whose projection on the t-space is G, and for any $(t,y)\in A$ let $\tilde{Q}(t,y)$ be a subset of the z^oz-space $\mathbb{R}^{r+1}$ with the property that $(z^o,z)\in\tilde{Q}(t,y)$, $z^{o'} \geq z^o$ implies $(z^{o'},z) \in \tilde{Q}(t,y)$. Take $\tilde{Q}(t,y)=\emptyset$ for $(t,y)\in\mathbb{R}^{\nu+s}-A$, and assume that for a.a.t the sets $\tilde{Q}(t,y)$ are convex and have property (K) with respect to y. Let $y_k(t), y(t), \xi_k(t), \xi(t), \lambda_k(t), \lambda(t), \eta_k(t)$, $t\in G$, $k=1,2,\dots$, be measurable functions with $\xi_k, \xi \in (L_p(G))^r$, $p \geq 1$, $\lambda_k, \lambda \in L_1(G)$, $\eta_k \in L_1(G)$, such that

$$y_k(t) \in A(t), \quad (\eta_k(t), \xi_k(t)) \in \tilde{Q}(t,y_k(t)), \quad \eta_k(t) \geq \lambda_k(t), \quad t\in G,$$

$$k=1,2,\dots, \quad -\infty < i = \liminf_k \int_G \eta_k(t)dt < +\infty.$$

Then there is a function $\eta(t)$, $t\in G$, $\eta \in L_1(G)$, such that

$$(\eta(t), \xi(t)) \in \tilde{Q}(t,y(t)), \quad \int_G \eta(t)dt \leq i.$$

(2.ii) <u>(An existence theorem for problem (2))</u>. Under condition (C'^{*}) assume that for a.a. $t \in G$ and all $y \in G$ the sets $\tilde{Q}(t,y)$ are convex and have property (K) with respect to y. Let us assume that (λ^{*}) for some $\psi(t) \geq 0$, $t \in G$, $\psi \in L_1(G)$, and constant $c \geq 0$ we have $f_0(t,y,u) \geq -\psi(t) - c|f(t,y,u)|$. Let Ω be a nonempty closed class $\Omega = \{(x,u)\}$ of elements $x \in X$ with $I[x,u]$ finite and (c) $S = \{x\}$ is (weakly) relatively compact in X. Assume that both M and L have the closure property, that at least one has the closed graph property, that M has the convergence in measure property, and L has the weak convergence property. Then the functional $I[x,u]$ in (2) has an absolute minimum in Ω.

A very general Tonelli-type theorem for problems (2) can now be stated as follows:

(2.iii) <u>(A Tonelli-type theorem existence theorem for problem (2))</u>. Same as before, where (λ^{*}) is replaced by (ϕ^{*}) there is a scalar function $\phi(\zeta)$, $0 \leq \zeta < +\infty$, bounded below, with $\phi(\zeta)/\zeta \to +\infty$ as $\zeta \to +\infty$, such that $f_0(t,y,u) \geq \phi(|f(t,y,u)|)$, and (c) is replaced by (c^{*}) any subclass $S^{*} = \{x\}^{*}$ of S with $\{Lx\}^{*}$ (weakly) relatively compact in $(L_1(G))^s$ is also (weakly) relatively compact in X.

As mentioned, problems (2) and (4) are equivalent, and the theorems above for problems of optimal control are equivalent to the theorems of Section 1 for extended problems of the calculus of variations. The proof of theorem (1.i) we shall sketch in Section 7 is also a proof of theorem (2.i). The property (K) we assume in (2.i) for the sets $\tilde{Q}(t,y)$ (with respect to y) corresponds to the assumption of lower semicontinuity in (y,z) for the function $F_0(t,y,z)$ (or $T(t,y,z)$).

3. Some more general problems of optimal control.

We may consider the problem of minimum with differential equations and constraints

$$I[x,u,v] = \int_G f_0(t,(Mx)(t),u(t))dt + \int_\Gamma g_0(\tau,(Kx)(\tau),v(\tau))d\mu, \quad x \in S,$$

$$(Lx)(t) = f(t,(Mx)(t),u(t)), \quad t \in G,$$

$(Jx)(\tau) = g(\tau,(Kx)(\tau),v(\tau)), \quad \tau \in \Gamma = \partial G,$

$(t,(Mx)(t)) \in A, \quad u(t) \in U(t,(Mx)(t)), \quad t \in G,$ (6)

$(\tau,(Kx)(\tau)) \in B, \quad v(\tau) \in V(\tau,(Kx)(\tau)), \quad \tau \in \Gamma,$

$f_o(\cdot,(Mx)(\cdot),u(\cdot)) \in L_1(G), \quad g_o(\cdot,(Kx)(\cdot),v(\cdot)) \in L_1(\Gamma),$

where G is a domain in the $\mathbb{R}^{\upsilon}$ space, $\upsilon \geq 1$, $t = (t^1,\dots,t^{\upsilon}) \in G$, and where the state variable x is thought of as an element of a subset S of a topological space $(X,\mathcal{T})$, possibly a Sobolev space X on G with its weak topology $\mathcal{T}$. Here $M: S \to (L_p(G))^s$, $L: S \to (L_p(G))^r$, $K: S \to (L_p(\Gamma))^{s'}$, $J: S \to (L_p(\Gamma))^{r'}$, $p \geq 1$, are given operators, not necessarily linear, and we write $y(t) = (Mx)(t), z(t) = (Lx)(t)$, $t \in G$, $\mathring{y}(\tau) = (Kx)(\tau)$, $\mathring{z}(\tau) = (Jx)(\tau)$, $\tau \in \Gamma$. Here $u(t) = (u^1,\dots,u^m)$, $t \in G$, is the control function on G, $v(\tau) = (v^1,\dots,v^{m'})$ the control function on Γ, and μ is the area measure on Γ. Finally, A is a subset of the ty-space whose projection on the t-space is G, and B is a subset of the $t\mathring{y}$-space whose projection on the τ-space is Γ. If we think of S as a subset of a Sobolev space X on G, involving derivatives of order up to a maximal order N, then M may be the set of all derivatives $D^{\alpha}x$, $0 \leq |\alpha| \leq N-1$, and K the set of the traces of the same derivatives on Γ. Then L and J may be differential operators on the same derivatives on G and their traces on Γ, of the orders N and N-1 at most.

As in Section 2 we may introduce sets $Q(t,y)$, $\tilde{Q}(t,y)$, $(t,y) \in A$ and $R(\tau,\mathring{y})$, $\tilde{R}(\tau,\mathring{y})$, $(\tau,\mathring{y}) \in B$, and two Lagrangians $T(t,y,z)$, $\mathring{T}(\tau,\mathring{y},\mathring{z})$, and problem (6) reduces to a problem of the calculus of variations on G and Γ with extended integrands. We omit the details and the existence theorems similar to those of Sections 1 and 2.

It may be that there is no differential equation on G and that $f_o = 0$, which is the case considered in a previous work by Fichera. Or it may occur that certain boundary values are given, that we have no differential equation on Γ, and $g_o = 0$, which is the case considered in Section 2.

4. The equivalence theorem

This theorem of analysis establishes conditions equivalent to weak convergence in L_1. The theorem can be stated for Lebesgue measures in $\mathbb{R}^\upsilon$, or for abstract measure spaces, finite, σ-finite, with or without atoms. We state it here in its simplest and typical form, namely for Lebesgue measures in a bounded region G in $\mathbb{R}^\upsilon$.

(4.i) (The equivalence theorem). Let $\{f(t), t \in G\}$ be a family of real valued L_1-integrable functions on the bounded region G in $\mathbb{R}^\upsilon$. The following statements are equivalent:

(a) The family $\{f\}$ is sequentially weakly relatively compact in $L_1(G)$.

(b) The family $\{f\}$ is equiabsolutely integrable in G.

(c) There is a constant M and a real valued function $\phi = \phi(\zeta)$, $0 \leq \zeta < + \infty$, bounded below, with $\phi(\zeta)/\zeta \to + \infty$ as $\zeta \to + \infty$, such that

$$\int_G \phi(|f(t)|)dt \leq M \quad \text{for all } f \in \{f\}.$$

(d) There is a real valued function $\psi = \psi(\zeta)$, $0 \leq \zeta < + \infty$, bounded below with $\psi(\zeta)/\zeta \to + \infty$ as $\zeta \to + \infty$, such that the family $\psi(|f(t)|)$, $t \in G$, $f \in \{f\}$, is equiabsolutely integrable in G.

In (c), (d) it is not restrictive to assume ϕ, ψ nonnegative strictly increasing, continuous, and convex in $[0,+\infty)$. Functions ϕ or ψ as above are often called Nagumo functions.

The equivalence of (a) and (b) was proved by Dunford and Pettis. The implication (b) → (c) was proved by De La Vallée Poussin. The implication (b) → (c) ∪ (d) was also proved by Candeloro and Pucci. The implication (c) → (b) was proved by Tonelli in particular cases, and then by Nagumo in the general case. The implication (d) → (c) is trivial. An elementary and direct proof of the whole statement for $\upsilon = 1$ has been given recently by Cesari and Pucci [3]. The proof can be extended to any $\upsilon \geq 1$. We refer to the last mentioned paper for the proof and for a number of references.

5. <u>Cesari's property (Q)</u>. Given sets $\tilde{Q}(t,y)$ as before, $(t,y) \in R^{\upsilon+s}$, $\tilde{Q}(t,y) \subset R^{r+1}$, we say that these sets have the property (Q) of upper semicontinuity with respect to y, provided for every (t_o,y_o) we have

$$\tilde{Q}(t_o,y_o) = \bigcap_{\delta > o} \text{cl co} \bigcup_{y \in N_\delta(y_o)} \tilde{Q}(t_o,y) . \qquad (7)$$

Equivalently, we may say that the set valued map $(t,y) \to \tilde{Q}(t,y)$ has this property. Sets having this property are necessarily closed and convex. This property (Q) is stronger than property (K) in the sense that property (Q) implies property (K) ([2], and [1], p. 293).

The typical use of property (Q) in proving lower closure theorems (lower semicontinuity) and thus in existence theorems in problems of optimization ([1], [2]) will be illustrated below.

Property (Q) has been used by many authors, for instance by Castaing and Valadier in theoretical questions in selection theorems, by Schuur in proving the existence of solutions of differential equations in Banach spaces with multivalued second members, by Olech, Lasota, and Baum in selection theorems, by Angell in proving the existence of solutions to nonlinear Volterra equations with delay and in nonlinear functional equations; and by Cesari and Hou in proving the existence of solution to nonlinear evolution equations (we refer to [1] for references). Cesari showed (cf. [1], p. 486) that property (Q) is the natural extension of the seminormality property used by Tonelli and McShane in the calculus of variations. Recently, Goodman [7] (cf. [1], p. 495) characterized property (Q) in terms of convex analysis. Finally, Suryanarayana (cf. [1], p. 500) has recently proved that property (Q) is a generalization of the concept of maximal monotonicity of Minty and Brezis. Indeed any maximal monotone map $z \to Q(z)$ in a real Hilbert space Z (thus, $z \in Z$, $Q(x) \subset Z$) has necessarily property (Q). More generally, Suryanarayana has proved that maximality with respect to any analytical property in a large class necessarily implies property (Q).

6. Growth property (ϕ) implies property (Q).

(6.i) Theorem ([2] and [1], p. 333). Let A be any set of points $x \in \mathbb{R}^h$, and for every $x \in A$ let $\tilde{Q}(x)$ denote a set of points $(y,z) = (y,z^1,\dots,z^r) \in \mathbb{R}^{1+r}$ such that (a) $(y,z) \in \tilde{Q}(x)$, $y \le y'$, implies $(y',z) \in \tilde{Q}(x)$. Let $\phi(\zeta)$, $0 \le \zeta < +\infty$, be a real valued function, bounded below, such that $\phi(\zeta)/\zeta \to +\infty$ as $\zeta \to +\infty$. For some $\bar{x} \in A$ let $N_\delta(\bar{x})$ be a neighborhood of $\bar{x}$ in A, and assume that (b) $(y,z) \in \tilde{Q}(x)$, $x \in N_\delta(\bar{x})$ implies $y \ge \phi(|z|)$. If the sets $\tilde{Q}(x)$ have property (K) at $\bar{x}$, and the set $\tilde{Q}(\bar{x})$ is convex, then the sets $\tilde{Q}(x)$ have property (Q) at $\bar{x}$.

This statement has a number of variants (cf. [1], p. 334) of which we mention here the following one:

(6.ii) Let A be any set of points $x \in \mathbb{R}^h$, and for every $x \in A$ let $Q(x)$ be a subset of the yz-space $\mathbb{R}^{1+r}$, let M denote the set $M = [(x,y,z) \mid x \in A, (y,z) \in Q(x)]$, let $T_o(x,y,z)$ be a real valued lower semicontinuous function on M, and let $\tilde{Q}(x)$ denote the set $\tilde{Q}(x) = [(v,y,z) \mid v \ge T_o(x,y,z), (y,z) \in Q(x)] \subset \mathbb{R}^{2+r}$. For some $\bar{x} \in A$ and neighborhood $N_\delta(\bar{x})$ of $\bar{x}$ in A assume that $x \in N_\delta(\bar{x})$, $(y,z) \in Q(x)$ implies $T_o(x,y,z) \ge \phi(|z|)$, $y \ge L$, where ϕ is a function as above and L a constant. If the sets $\tilde{Q}(x)$ have property (K) at $\bar{x}$ and the set $\tilde{Q}(\bar{x})$ is convex, then the sets $\tilde{Q}(x)$ have property (Q) at $\bar{x}$ (cf. [1], p. 334, (10.5.ii), second part, $\mu = 1$).

7. Sketch of proofs.

(a) Concerning existence theorems (1.ii) and (2.ii) we have already pointed out that the property (K) of the sets $\tilde{Q}(t,y)$ with respect to y in (1.ii) is the necessary and sufficient condition for the Lagrangian $F_o(t,y,z)$ to be lower semicontinuous in (y,z). Thus, it is enough to prove (1.ii). In the notations of Section 1 we denote by $\tilde{Q}(t,y)$ the sets $\tilde{Q}(t,y) = [(z^o,z) \mid \infty > z^o \ge F_o(t,y,z)]$.
Let $i = \mathrm{Inf}(I[x], x \in S)$, $-\infty \le i < +\infty$. There is a minimizing sequence $[x_k]$ such that $I[x_k] \to i$ as $k \to \infty$, $x_k \in S$, and then

$$I[x_k] = \int_G F_o(t,y_k(t),\xi_k(t))dt, \quad (Lx_k)(t) = \xi_k(t) \in Q(t,y_k(t)),$$
$$(Mx_k)(t) = y_k(t), \quad (t,y_k(t)) \in A, \quad \eta_k(t) = F_o(t,y_k(t),\xi_k(t)), \qquad (8)$$
$$(\eta_k(t),\xi_k(t)) \in \tilde{Q}(t,y_k(t)), \quad t \in G, \ k=1,2,\ldots .$$

Since S is sequentially relatively weakly compact in X, there is a subsequence, say still $[k]$ for the sake of simplicity, such that $x_k \to x \in X$ weakly in X.

Since L has the weak convergence property and M has the convergence in measure property, there is a subsequence, say still $[k]$, such that $Lx_k = \xi_k \to \xi$ weakly in $(L_1(G))^r$ and $Mx_k = y_k \to y$ in measure in G. At least one of the operators L and M has the closed graph property, hence $x \in S$. Both L and M have the closure property, hence $\xi = Lx$, $y = Mx$. Since $\xi_k \to \xi$ weakly in L_1, then $\|\xi_k\|_1 \leq C$ for some constant C. Thus, $\eta_k(t) = F_o(t,y_k(t),\xi_k(t)) \geq \psi(t) - c|\xi_k(t)|$, and $I[x_k] = \int_G \eta_k dt \geq -\|\psi\|_1 -cC$; hence $I[x_k]$ is bounded below, and i is finite.

(b) To complete the existence proof we need, as usual, the lower semicontinuity theorem (1.i) or the essentially equivalent lower closure theorem (2.i). We sketch the basic proof here.

We note that there we can take $\lambda_k(t) = -\psi(t) - c|\xi_k(t)|$. Since ξ_k converges weakly in $(L_1(G))^r$, by the equivalence theorem (implications (a) $\to$ (b), (b) $\to$ (a)) we derive that $|\xi_k|$ contains a subsequence, say still $[k]$, which converges weakly in $L_1(G)$, and hence λ_k is also weakly convergent in $L_1(G)$ toward a scalar function, say $\lambda(t)$, $t \in G$, $\lambda \in L_1(G)$. These are the functions λ_k, λ of theorems (1.i) and (2.i).

If $j_k = I[x_k]$ then $j_k \to i$ as $k \to \infty$, and if $\delta_s = \max [|j_k-i|, k \geq s+1]$, then $\delta_s \to 0$ as $s \to \infty$. Here $\xi_k \to \xi$ weakly in $(L_1(G))^r$, and again by the equivalence theorem, implication (a) $\to$ (d), there is a scalar function $\phi(\zeta)$, $0 \leq \zeta < +\infty$, nonnegative, continuous, increasing, convex, with $\phi(\zeta)/\zeta \to +\infty$ as $\zeta \to +\infty$, and such that the sequence $[\rho_k(t) = \phi(|\xi_k(t)|)$, $t \in G$, $k=1,2,\ldots]$ is equiabsolutely integrable, hence there is a sequence, say still $[k]$, such that $\rho_k(t) \to \rho(t)$ weakly in $L_1(G)$, and $\rho(t) \geq 0$ is an element of $L_1(G)$.

Now, for any $s = 1,2,3,\ldots$, the sequence ρ_{s+k}, λ_{s+k}, ξ_{s+k}, $k=1,2,\ldots$, converges weakly to ρ,λ,ξ in $(L_1(G))^{r+2}$. By the Banach-Saks-Mazur theorem (cf., e.g. [1], p. 325) there is a set of real numbers $c_{Nk}^{(s)} \geq 0$, $k = 1,\dots,N$, $N = 1,2,\dots$, with $\Sigma_k c_{Nk}^{(s)} = 1$, where Σ_k ranges over $k = 1,\dots,N$, and such that, if

$$\rho_N^{(s)}(t) = \Sigma_k c_{Nk}^{(s)} \rho_{s+k}(t),\ \lambda_N^{(s)}(t) = \Sigma_k c_{Nk}^{(s)} \lambda_{s+k}(t),\ \xi_N^{(s)}(t) = \Sigma_k c_{Nk}^{(s)} \xi_{s+k}(t),$$

$t \in G$, $N = 1,2,\dots$, then $(\rho_N^{(s)},\lambda_N^{(s)},\xi_N^{(s)}) \to (\rho,\lambda,\xi)$ strongly in $(L_1(G))^{r+2}$, and this is true for every $s = 1,2,\dots$. Thus, there is a sequence of integers N_ℓ such that $(\rho_{N_\ell}^{(s)}(t),\lambda_{N_\ell}^{(s)}(t),\xi_{N_\ell}^{(s)}(t)) \to (\rho(t),\lambda(t),\xi(t))$ as $\ell \to \infty$, for $t \in G$ (a.e.). Let

$$\eta_N(t) = \Sigma_k c_{Nk}^{(s)} \eta_{s+k}(t),\ t \in G,$$

and note that

$$\rho_k(t) = \phi(|\xi_k(t)|),\quad \eta_k(t) \geq \lambda_k(t),\ t \in G,\ \int_G \eta_k(t)dt = j_k,\quad k=1,2,\dots,$$

and hence

$$\eta_N^{(s)}(t) \geq \lambda_N^{(s)}(t),\quad t \in G,\quad i - \delta_s \leq \int_G \eta_N^{(s)}(t)dt \leq i+\delta_s, \tag{9}$$

with $\delta_s \to 0$ as $s \to \infty$. For $N = N_\ell(s)$ and $\ell \to \infty$, and by Fatou's lemma (cf. [1], p. 301), relations (9) imply

$$\eta^{(s)}(t) = \liminf_{\ell\to\infty} \eta_{N_\ell}^{(s)}(t) \geq \lambda(t),\quad t \in G \text{ (a.e.)},$$

$$\int_G \eta^{(s)} dt \leq \liminf_{\ell\to\infty} \int_G \eta_{N_\ell}^{(s)}(t)dt \leq i + \delta_s,$$

and for $\eta(t) = \liminf_{s\to\infty} \eta^{(s)}(t)$, we also have $\eta(t) \geq \lambda(t)$, $t \in G$, $\int_G \eta(t)dt \leq i$.

For a.a. $t_o \in G$, and $\varepsilon > 0$, there is now an s_o such that, for $s \geq s_o$ we have $\eta_N^{(s)}(t_o) \geq \lambda(t_o)-1$ and $|y_s(t_o)-y_o| \leq \varepsilon$ where $y_o = y(t_o)$. For $s \geq s_o$ we certainly have

$$(\eta_{s+k}(t_o),\ \xi_{s+k}(t_o)) \in \tilde{Q}(t_o,y_{s+k}(t_o)),\ |y_{s+k}(t_o)-y_o| \leq \varepsilon, k=1,2,\dots. \tag{10}$$

We need now the sets

$$\tilde{Q}'(t_o,y) = [(\eta,z)\,|\,\eta \geq \lambda(t_o)-1, (\eta,z) \in \tilde{Q}(t_o,y)] \subset \mathbb{R}^{r+1},\ (t_o,y) \in A,$$

$$\tilde{Q}'^{*}(t_o,y) = [(v,\eta,z)|v \geq \phi(|z|), \eta \geq \lambda(t_o)-1, (\eta,z) \in \tilde{Q}(t_o,y)] \subset \mathbb{R}^{r+2},$$

$$(t_o,y) \in A.$$

The sets $\tilde{Q}(t_o,y)$ have property (K) with respect to y since they are the intersection of the sets $\tilde{Q}(t_o,y)$, which have this property, with the fixed set $[(\eta,z)|\eta \geq \lambda(t_o)-1,\ z \in \mathbb{R}^r]$. The sets $\tilde{Q}'^{*}(t_o,y)$ are convex since the sets $\tilde{\tilde{Q}}(t_o,y)$ and $\tilde{Q}'(t_o,y)$ are convex and ϕ is convex. Again for t_o fixed, we apply (6.ii) with the variables x,y,z replaced by y,η, z, with the sets Q(x) in the yz-space replaced by the sets $\tilde{Q}'(t_o,y)$ in the ηz-space, with the sets $\tilde{Q}(x)$ in the vyz-space replaced by the sets $\tilde{Q}'^{*}(t_o,y)$ in the vηz-space, with $T_o(x,y,z)$ replaced by $T_o(y,\eta,z) = \phi(|z|)$, continuous, and since the sets $\tilde{Q}'(t_o,y)$ have property (K) with respect to y, and the sets $\tilde{Q}'^{*}(t_o,y)$ are convex, we derive from (6.ii) that the sets $\tilde{Q}'^{*}(t_o,y)$ have property (Q) with respect to y, and this holds for a.a. $t_o \in G$.

From (10) we have now, for $s \geq s_o$,

$$(\rho_{s+k}(t_o),\eta_{s+k}(t_o),\xi_{s+k}(t_o)) \in \tilde{Q}'^{*}(t_o,y_{s+k}(t_o)), \quad |y_{s+k}(t_o)-y_o| \leq \varepsilon,$$

and hence

$$(\Sigma_k c_{Nk}^{(s)} \rho_{s+k}(t_o),\ \Sigma_k c_{Nk}^{(s)} \eta_{s+k}(t_o), \Sigma_k c_{Nk}^{(s)} \xi_{s+k}(t_o) \varepsilon \text{ co } \tilde{Q}'^{*}(t_o,y_o,\varepsilon),$$

$$s \geq s_o,$$

where $Q'^{*}(t_o,y_o,\varepsilon)$ denotes the union of all $\tilde{Q}'^{*}(t_o,y)$ for $|y-y_o| \leq \varepsilon$. For $N = N_\ell$ and $\ell \to \infty$ the points in the first member of this relation form a sequence possessing $(\rho(t_o),\eta^{(s)}(t_o),\xi(t_o))$ as an element of accumulation in $\mathbb{R}^{n+2}$; hence

$$(\rho(t_o),\eta^{(s)}(t_o),\xi(t_o)) \in \text{cl co } \tilde{Q}'^{*}(t_o,y_o,\varepsilon),\ s \geq s_o.$$

Since $\eta(t_o) = \liminf \eta^{(s)}(t_o)$ as $s \to \infty$, we derive that

$$(\rho(t_o),\ \eta(t_o),\ \xi(t_o)) \in \text{cl co } \tilde{Q}'^{*}(t_o,y_o,\varepsilon).$$

Since $\varepsilon > 0$ is arbitrary, by property (Q) we derive that

$$(\rho(t_o),\eta(t_o),\ \xi(t_o)) \in \bigcap_{\varepsilon > 0} \text{cl co } \tilde{Q}'^{*}(t_o,y_o,\varepsilon) = \tilde{Q}'^{*}(t_o,y_o).$$

By the definition of $\tilde{Q}$ we have then

$$(\eta(t_o),\ \xi(t_o)) \in \tilde{Q}(t_o, y(t_o)) \tag{11}$$

and this holds for a.a. $t_o \in G$. Hence $\eta(t) \geq F_o(t,y(t),\xi(t))$, $t \in G$ (a.e.), and

$$\begin{aligned} i \geq \int_G \eta(t)dt \geq \int_G F_o(t,y(t),\xi(t))dt,\ \xi(t) = (Lx)(t), \\ y(t) = (Mx)(t), \end{aligned} \tag{12}$$

where, at this point, F_o is measurable with Lebesgue integral finite or $-\infty$. This completes the lower semicontinuity argument.

(c) The argument for the existence theorem continues now with the usual remark that $\eta(t) \geq F_o(t,y(t), \xi(t)) \geq -\psi(t) - c|\xi(t)|$, hence F_o is L-integrable in G, and $I[x] = \int_G F_o dt \leq i$. Since S is closed, $x \in S$, hence $I[x] \geq i$, and then $I[x] = i$. This completes the proof of the existence theorems (1.ii) and (2.ii).

(d) The proof above has drastic simplifications if it is known that the original sets $\tilde{Q}(t,y)$ have property (Q) with respect to y, since then there is no need to construct the auxiliary sets $\tilde{Q}'^{*}(t,y)$. We refer to ([1], p. 345, Second proof) for the details. This is what happens in the situation depicted by the Tonelli-type theorems (1.iii) and (2.iii).

8. Criteria of the F-, G-, H-types.

In applying existence theorems of Section 2 to given problems for which the functions f_o and f are given, the property (K) for the sets $\tilde{Q}(t,y)$ may not be easy to verify. Even more difficult may be the verification that the corresponding Lagrangian $T(t,y,z)$ is lower semicontinuous in y,z because T, though defined by (3), may not be known explicitly. The following conditions, analytical in character and easy to verify, have a practical significance. These conditions are discussed in detail in [1] (Chapter 13) for the one dimensional case, but hold for multidimensional problems as well. In each of the conditions, we deal with a sequence x_k, say a minimizing sequence, corresponding sequence $y_k(t) = (Mx_k)(t)$, $t \in G$, $k=1,2,\dots$, with $y_k \to y$ as $k \to \infty$, and the differences

$$\delta_k(t) = f(t,y_k(t),u_k(t)) - f(t,y(t),u_k(k)),$$

$$\delta_{ok}(t) = f_o(t,y_k(t),u_k(t)) - f_o(t,y(t),u(t)), \quad t \in G, \quad k=1,2,\ldots .$$

Under each condition certain requirements are made, and each of them implies a property D, analytical in character, which in turns guarantes a weak property (Q) for the sets $\tilde{Q}(t,y_k(t))$. This suffices for the conclusion of the simplified proof mentioned at the end of Section 7. We refer to [1] (Chapter 13) for the details, for proofs, and for the condition (D) mentioned above. We state here only a few of the criteria under consideration. For these conditions, some of which were first proposed by Rothe [9], we refer to Cesari and Suryanarayana [4]. Condition (D) was proved by Angell [7] to be relevant in questions of approximation in the calculus of variations.

<u>Lipschitz-type condition F_p: geometric viewpoint</u>. For $1 \le p < +\infty$, $y, y_K \in (L_p(G))^s$, $\|y_k - y\|_p \to 0$, and $|\delta_k(t)|$, $|\delta_{ok}(t)| \le F_k(t)h(|y_k(t) - y(t)|)$, $t \in G$, $k=1,2,\ldots$, where $h(\zeta) \ge 0$, $0 \le \zeta < +\infty$, is a given monotone nondecreasing function with $h(+0) = 0$, and $h(\zeta) \le c|\zeta|^\gamma$ for all $\zeta \ge \zeta_o > 0$, $c \ge 0$, $0 < \gamma \le p$, (c, γ, ζ_o given constants), and $F_k(t) \ge 0$, $t \in G$, $F_k \in L_{p'}(G)$, given functions with $p' = p/(p-\gamma)$, ($p' = \infty$ if $\gamma = p$), and $\|F_k\|_{p'} \le M$, a given constant.

<u>Lipschitz-type condition F_p: functional viewpoint</u>. For $1 \le p < +\infty$, let $c, \gamma, \zeta_o, p', h$ as in (F_p) above, and there is a function $F(t,u) \ge 0$ such that $\int_G |F(t,u(t))^{p'} dt \le M$ for all admissible control functions $u(t)$, $t \in G$, and

$$|f(t,y_1,u(t) - f(t,y_2,u(t)| \le F(t,u(t))h(|y_1 - y_2|),$$

and the same holds for f_o.

For variants of these conditions, and condition F_∞, we refer to [1] (Chapter 13). For the remaining conditions here we mention only the geometric viewpoint.

Growth-type condition G_{pq}. (i) There exists a continuous function $\phi(u)$, $\phi: \mathbb{R}^m \to \mathbb{R}^m$, with $|\phi(u)| \to \infty$, such that for $1 \le p, q < +\infty$, $y, y_k \in (L_p(G))^s$, u_k measurable, $\phi(u_k) \in (L_q(G))^m$, $\|y\|_p, \|y_k\|_p \le L_o$, $\|\phi(u)\|_q \le L$ (L_o, L given constants), $y_k(t) \to y(t)$ in measure in G as $k \to \infty$, and (ii) there are constants c, c', α, β with $c, c' \ge 0$, $0 < \alpha \le p$, $0 < \beta \le q$, and a function $\psi(t) \ge 0$, $t \in G$, $\psi \in L_1(G)$, such that, for all (t,y,u), $(t,z,u) \in M$ we have

$$|\delta_k(t)| \, , \, |\delta_{ok}(t)| \le \psi(t) + c(|y_k(t)|^{p-\alpha} + |y(t)|^{p-\alpha}) + c'|\phi(u_k(t)|^{q-\beta}$$

We refer to [1] for conditions $G_{\infty q}$, $G_{p\infty}$, $G_{\infty,\infty}$.

Growth-type conditions H_q. (i) For $1 \le q < \infty$, y, y_k measurable, $y_k(t) \to y(t)$ in measure in G as $k \to \infty$, $u_k \in (L_q(G))^m$, $\|u_k\|_q \le L$, a constant, and (ii) there are other constants c', β , $0 < \beta \le q$, and a function $\psi(t) \ge 0$, $t \in G$, $\psi \in L_1(G)$, such that for all (t,y,u), $(t,z,u) \in M$, we have

$$|f(t,y,u) - f(t,z,u)| \le \psi(t) + c' |u|^{q-\beta}$$

and the same holds for f_o.
We refer to [1] for condition H_∞.

Addendum (Oct. 1983).

This lecture, as delivered in Bologna, contained also a presentation of work of C. Vinti and his associates at the University of Perugia in the last ten years (M. Boni, P. Brandi, C. Gori, P. Pucci, M. Ragni, A. Salvadori et al.) on the existence and properties of the Weierstrass integrals of the calculus of variations, parametric and nonparametric, over curves and surfaces. Two detailed expositions of this material by C. Vinti are now in process of publication:

C. VINTI, The integrals of the calculus of variations in the sense of Weierstrass-Burkill-Cesari. "Parametric Optimization and Related

Topics" (L. Cesari, R. Goor, E. Silverman, C. Vinti). Springer-Verlag Lecture Notes in Math. (to appear).

C. VINTI, Nonlinear integration and Weierstrass integral over a manifold: connections with theorems on martingales. Journal Optimization Theory and Applications, Vol. 41, No. 1, September 1983, pp. 213-235.

With the agreement of the Organizing Committee, we refer to these publications for the material above.

Also, we refer to

L. CESARI, Existence of solutions and existence of optimal solutions. A lecture at the Intern. Conference in S. Margherita Ligure, Nov. 30-Dec. 4,1981. "Mathematical Theories of Optimization", Springer-Verlag Lecture Notes in Math. No. 979 (Cecconi and Zolezzi eds.), 1983, pp. 88-107;

L. CESARI, Nonlinear analysis. A lecture at the 12th Congress of the Unione Matematica Italiana, Perugia, Sept. 2-7, 1983. To appear in Bollettino Un. Mat. Ital., Section A;

for material concerning existence of solutions to boundary value problems for nonlinear ordinary and partial differential equations.

References

[1] L. CESARI, Optimization - Theory and Applications. Problems with Ordinary Differential Equations. Springer Verlag 1983, xiv+542.

[2] L. CESARI, Existence theorems for weak and usual optimal solutions in Lagrange problems with unilateral constraints. I and II. Trans. Amer. Math. Soc. 124, 1966, 369-412, 413-429.

[3] L. CESARI and P. PUCCI, An elementary proof of an equivalence theorem relevant in the theory of optimization. Journ. Optimization Theory Appl., to appear.

[4] L. CESARI and M.B. SURYANARAYANA, Closure theorems without seminormality conditions. Journ. Optimization Theory Appl. 15, 1975, 441-465.

[5] L. CESARI and M.B. SURYANARAYANA, On recent existence theorems in the theory of optimization. Journ. Optimization Theory Appl. 31, 1980, 197-415.

[6] L. CESARI and M.B. SURYANARAYANA, Upper semicontinuity properties of set valued functions. Nonlinear Analysis 4, 1980, 639-656.

[7] T.S. ANGELL, A note on approximation of optimal solutions of free problems of the calculus of variations. Rend. Circ. Mat. Palermo (2) 28, 1979, 258-272.

[8] G.S. GOODMAN, The duality of convex functions and Cesari's property (Q). Journ. Optimization Theory Appl. 19, 1976, 17-23.

[9] E.H. ROTHE, An existence theorem in the calculus of variations, Arch. Rat. Mech. Anal. 21, 1966, 151-162.

BEHAVIOUR OF SOLUTIONS OF NONLINEAR ALTERNATIVE PROBLEMS UNDER PERTURBATIONS OF THE LINEAR PART WITH RANK CHANGE

Heinz W. Engl

Institut für Mathematik
Johannes-Kepler-Universität
A-4040 Linz, Austria

Abstract:

This paper is concerned with the behaviour of solutions of nonlinear operator equations with non-invertible linear part under perturbations of the operators involved. Even in the linear case, the solutions need not change continously if the linear part is perturbed in such a way that its rank changes. We prove a continuous-dependence result for the nonlinear problem and illustrate it with an example involving the behaviour of periodic solutions of nonlinear differential equations at resonance under perturbations of the differential operator.

1. Introduction and Preliminaries

In this paper, we study the question how a solution of a nonlinear alternative problem behaves if the linear part of the equation is perturbed in such a way that its rank changes. More precisely, let X be a real Banach space and for $\varepsilon \geq 0$, let L_ε be a linear operator, N_ε a nonlinear operator. We assume that dim $N(L_0) > 0$, so that for $\varepsilon = 0$ the problem

$$(1.1) \qquad L_\varepsilon x = N_\varepsilon(x)$$

is really a "nonlinear alternative problem" or a "problem at resonance" in the terminology of Cesari [2] . We will try to investigate what happens to particular solutions of (1.1) if ε increases from 0 to nonzero values, where we assume that

(1.2) $$\dim N(L_\varepsilon) \neq \dim N(L_0) \text{ for } \varepsilon > 0.$$

Our motivation for studying this question is at least twofold. First, it is a question of stability of solutions of (1.1) for $\varepsilon = 0$ under data changes if one views the operator L_0 as part of the data. Since L_0 is assumed to be singular, even small changes in L_0 will be likely to change the rank. Our results will give some information what kinds of changes in L_0 will be harmless in the sense that the solution of (1.1) is continuous in ε at $\varepsilon = 0$, which means that small changes in L_0 will not alter the solution drastically.

Secondly, the problem addressed here might be important if one tries to design a numerical method for solving nonlinear alternative problems in the spirit of "continuation methods" (see e.g. [7]) in such a way that the original problem is imbedded into a parameterized family of problems in such a way that the dimension of the nullspaces of the linear parts increases from 0 to the value for the original problem at discrete points in the parameter interval.

In order to continue the solution across these points, one needs a solution to our problem studied here.
We have to admit, however, that the results presented here can be considered only as a first step towards a real contribution to these problems, since the requirements especially on the kinds of admissible nonlinearities are rather stringent.

Before addressing the nonlinear problem, a few remarks concerning the corresponding linear problem are appropriate, since even the linear problem is nontrivial and since the linear results are used in obtaining a solution for the nonlinear problem. The linear problem consists of studying the behaviour of solutions of

(1.3) $$L_\varepsilon x = f_\varepsilon \quad ,$$

where L_ε is as above, at $\varepsilon = 0$ under the assumption that (1.2) holds. The following trivial example shows that even in the finite-dimensional case one cannot expect continuity of any solutions without additional requirements.

Example 1.1:

For $\varepsilon \geq 0$, let $L_\varepsilon : \mathbb{R}^2 \to \mathbb{R}^2$ be represented by the matrix

$\begin{pmatrix} 1+\varepsilon & \varepsilon \\ \varepsilon & \varepsilon^2 \end{pmatrix}$, $f_\varepsilon := \begin{pmatrix} 1 \\ 0 \end{pmatrix}$. For all $\varepsilon \geq 0$, the equation (1.3) is solvable; for $\varepsilon > 0$, the unique solution is $\begin{pmatrix} \varepsilon^{-1} \\ -\varepsilon^{-2} \end{pmatrix}$, which does not converge to anything, let alone a solution of $L_0 x = f_0$, as $\varepsilon \to 0$.

The linear problem has been studied extensively in [4] and been applied to linear integral equations arising from electrostatic and magnetostatic boundary value problems there. The nature of these problems was such that the linear operators involved could be assumed to be bounded. Here, we have to admit that the linear operators are densely defined and closed, but unbounded, since one of the major fields of applications of nonlinear alternative problems is in treating semilinear differential equations. There, either the linear operators involved are unbounded or the underlying spaces have to be chosen in such a way that the operators are bounded, but do not map the spaces into themselves. Since for the linear theory it is important that the operators map a Banach space into itself, we have to extend the results of [4] to the case of closed, not necessarily bounded linear operators. This is done in Section 2.

For later use, we review some aspects of the generalized inverse of a linear operator on a Banach space. For a comprehensive treatment of generalized inverses in the framework of operator theory see [5]. It should be noted that L. Cesari used generalized inverses under the name of "partial inverse" at least as early as 1963 [1].

Let X be a real Banach space, T a closed linear operator with dense domain $D(T) \subseteq X$ and range $R(T) \subseteq X$. Suppose that the nullspace $N(T)$ has a topological complement M in X and that $\overline{R(T)}$ has a topological complement S in X. By P and Q we denote the projectors onto $N(T)$ and $\overline{R(T)}$ induced by the decompositions $X = N(T) \oplus M$ and $X = \overline{R(T)} \oplus S$, respectively. The "generalized inverse" $T^\dagger$ of T (with respect to P and Q) is defined as the unique linear extension of $(T|_{D(T)\cap M})^{-1}$ (defined on $R(T)$) to $D(T^\dagger) := R(T) \dot{+} S$ such that $N(T^\dagger) = S$. Note that $T^\dagger$ depends on the choice of M and S, or equivalently, of P and Q. Whenever we want to stress this dependence, we write $T^\dagger_{P,Q}$. It is well-known that $T^\dagger$ is bounded iff $R(T)$ is closed.

$T^\dagger_{P,Q}$ can be characterized by the equations

$$(1.4) \qquad T^\dagger T T^\dagger = T^\dagger \qquad \text{on } D(T^\dagger),$$

$$(1.5) \qquad T T^\dagger T = T \qquad \text{on } D(T),$$

$$(1.6) \qquad T T^\dagger = Q \qquad \text{on } D(T^\dagger),$$

$$(1.7) \qquad T^\dagger T = (I-P) \qquad \text{on } D(T).$$

2. The Linear Problem

As indicated in the introduction, we generalize the results of [4] concerning the behaviour of solutions of (1.3) as $\varepsilon \to 0$ under the assumption that (1.2) holds to the case of unbounded operators L_ε. To be specific, let X be a real Banach space, D a dense subset of X. For all $\varepsilon \geq 0$, let $L_\varepsilon : D \to X$ be a closed linear operator with closed range and $f_\varepsilon \in R(L_\varepsilon)$. We assume that

$$(2.1) \qquad \dim N(L_\varepsilon) = n \text{ for all } \varepsilon > 0$$

with $n \in \mathbb{N}_0$ independent of ε. Furthermore, we assume that

(2.2) $$\dim N(L_o) = m$$

with $m \in \mathbb{N}_o$ (not necessarily = n) and that

(2.3) $$X = N(L_o) \oplus R(L_o) .$$

By P we denote the projector onto $N(L_o)$ induced by (2.3). Finally, we assume that for all $\varepsilon > 0$,

(2.4) $$K_\varepsilon := L_\varepsilon - L_o$$

is continuous. We identify K_ε with its unique extension to a bounded linear operator on all of X. It might be considered restrictive to assume that the domain of L_ε is independent of ε and that $L_\varepsilon - L_o$ is continuous. But these assumptions are reasonable for the classes of operators we have in mind: If the L_ε are differential operators of the same order, D may be the set of continuous functions for which the derivative of this order exists and is continuous, thus D will not depend on ε. The assumption of continuity of $L_\varepsilon - L_o$ may then be fulfilled if $L_\varepsilon - L_o$ does not contain derivatives, which is of course restrictive. An approach which would permit $L_\varepsilon - L_o$ to contain derivatives would require to consider the L_ε as operators between different spaces (where the domain space will be continuously imbedded into the range space); although it might be worthwile to pursue this line, this will not be done here.

We now assume that

(2.5) $$L_\varepsilon \to L_o,\ f_\varepsilon \to f_o \quad \text{as} \quad \varepsilon \to 0,$$

where "$L_\varepsilon \to L_o$" means "$\|L_\varepsilon - L_o\| \to 0$", which makes sense because of our assumption that K_ε is continuous.

The proof of the main result of this section proceeds in a way similiar to the proof of Theorem 2.9 in [4], only a few steps are different. Therefore we only outline the proof and refer to [4] for details not provided here. It should be noted that unfortunately [4] contains a few misprints: nearly all Landau symbols "O" there should be replaced by "o"; this should be clear from the context, however.

Lemma 2.1: L_o+P is a closed bijection from D to X. The bounded linear operator

$$(2.6) \qquad L_o^\dagger := (L_o+P)^{-1}(I-P)$$

from X onto D is the generalized inverse of L_o with respect to the projectors P and I-P.

Proof: The closedness of L_o+P follows from the continuity of P (which in turn follows from (2.3)) and the closedness of L_o.

Let $x \in D$ be such that $(L_o+P)x = 0$. Because of (2.3), there are unique $x_N \in N(L_o)$ and $x_R \in R(L_o)$ with $x = x_N+x_R$. Then $0= (L_o+P)x=L_ox_R+x_N$. Again because of (2.3) this implies $x_N = 0$ and $L_ox_R = 0$, i.e., $x_R \in N(L_o)$. Since $x_R \in R(L_o)$, we have $x_R = 0$ by (2.3) and thus $x = 0$. Thus L_o+P is injective.

Now, let $y = y_N+y_R$ be arbitrary, but fixed (where $y_N \in N(L_o)$, $y_R \in R(L_o)$). There is an $x_R \in D$ with $L_ox_R = y_R$; since $L_oPx_R = 0$, we have $y = L_o(I-P)x_R+$ $+y_N = L_o(y_N+(I-P)x_R)+P(y_N+(I-P)x_R) \in R(L_o+P)$.

Thus, L_o+P is also surjective.

By the Closed Graph Theorem, $(L_o+P)^{-1}:X \to D \subseteq X$ is bounded. To show that $(L_o+P)^{-1}(I-P)$ is in fact $L^\dagger_{oP,I-P}$, we could check the equations (1.4) through (1.7). We can also use the following argument:

By definition, $L^\dagger_{oP,I-P}$ is the unique linear extension of $(L_o|_{D\cap R(L_o)})^{-1}$ (definied on $R(L_o)$) to all of X such that $N(L^\dagger_{oP,I-P}) = R(P) = N(L_o)$. We claim that this linear extension is $(L_o+P)^{-1}(I-P)$. Indeed, $N((L_o+P)^{-1}(I-P))= R(P)$. Thus, it suffices to show that for $y \in R(L_o)$, $(L_o|_{D\cap R(L_o)})^{-1}y = (L_o+P)^{-1}(I-P)y$. Let $y \in R(L_o)$. Then $(L_o+P)^{-1}(I-P)y =$ $= (L_o+P)^{-1}y =: z \in D$. This element z is characterized by $L_oz+Pz = y$.

Since $L_o z-y \in R(L_o)$, we have $Pz = 0$ because of (2.3). Thus, $L_o z = y$ and $z \in D \cap R(L_o)$, i.e., $z = (L_o|_{D\cap R(L_o)})^{-1}y$. Together with the definition of z, this completes the proof. □

Now, let for all $\varepsilon \geq 0$, $F_\varepsilon : D \to R(L_o)$ and $M_\varepsilon : X \to R(L_o) \cap D$ be defined by

$$(2.7) \qquad F_\varepsilon := (I-P)L_\varepsilon \ , \quad M_\varepsilon = L_o^\dagger (L_o - L_\varepsilon),$$

where $L_o^\dagger$ is defined as in (2.6) and $L_o - L_\varepsilon$ is meant to be the unique bounded extension to all of X. With these notations we have:

Lemma 2.2: For all sufficiently small $\varepsilon > 0$, $I-M_\varepsilon$ has a bounded inverse and

$$(2.8) \qquad F_\varepsilon = L_o(I-M_\varepsilon)$$

holds on D. In particular,

$$(2.9) \qquad \begin{aligned} N(F_\varepsilon) &= (I-M_\varepsilon)^{-1} N(L_o) \\ R(F_\varepsilon) &= R(L_o) \end{aligned}$$

hold for sufficiently small $\varepsilon > 0$.

Proof: Because of Lemma 2.1, $L_o^\dagger$ is bounded; together with (2.5) this implies $\|M_\varepsilon\| < 1$ for sufficiently small $\varepsilon > 0$. For those ε, $I-M_\varepsilon$ is boundedly invertible, and we have (with convergence in the norm)

$$(2.10) \qquad (I-M_\varepsilon)^{-1} = \sum_{i=o}^{\infty} M_\varepsilon^i .$$

With the help of (2.10) and the equations (1.4)-(1.7) for $L_o^\dagger$ we verify (2.8) as in Lemma 2.5 of [4], taking into account that $R(M_\varepsilon) \subseteq D$. Then (2.9) follows immediately. □

For $\varepsilon > 0$ as small as needed in Lemma 2.2, let $F_\varepsilon^\dagger : X \to R(L_o)$ be defined by

(2.11) $$F_\varepsilon^\dagger := (I-M_\varepsilon)^{-1} L_o^\dagger ,$$

where $L_o^\dagger$ is as in (2.6).

<u>Lemma 2.3</u>: For ε sufficiently small, $P_\varepsilon := (I-M_\varepsilon)^{-1}P$ is a projector onto $N(F_\varepsilon)$; $F_\varepsilon^\dagger$ is the generalized inverse of F_ε with respect to the projectors $P_\varepsilon: X \to N(F_\varepsilon)$, $I-P: X \to R(F_\varepsilon)$.

<u>Proof</u>: Because of (2.9), $R(P_\varepsilon) = N(F_\varepsilon)$. It suffices to prove that for $x \in N(F_\varepsilon)$, $P_\varepsilon x = x$. Let $x \in N(F_\varepsilon)$. Because of (2.9), $(I-M_\varepsilon)x \in N(L_o)$, i.e., there is an $n \in N(L_o)$ with $x = M_\varepsilon x+n$ and (since $R(M_\varepsilon) \subseteq R(L_o)$) $Px = n$. Thus, $Px = x-M_\varepsilon x$ and therefore $P_\varepsilon x = (I-M_\varepsilon)^{-1}Px = x$. Thus, P_ε is indeed a projector onto $N(F_\varepsilon)$. As in [4], we show that $F_\varepsilon^\dagger$ fulfills (1.4)-(1.7) with $(I-P)$ instead of Q and P_ε instead of P there. The only additional observation one has to make is that $R(F_\varepsilon^\dagger) \subseteq D$, which holds because of (2.10) and the facts that $R(L_o^\dagger) \subseteq D$ and $R(M_\varepsilon) \subseteq D$. □

Now, let

(2.12) $$\phi_o := L_o^\dagger f_o$$

and for $\varepsilon > 0$ sufficiently small für Lemma 2.2 to hold

(2.13) $$\phi_\varepsilon := F_\varepsilon^\dagger f_\varepsilon .$$

<u>Lemma 2.4</u>: $\phi_\varepsilon \to \phi_o$ as $\varepsilon \to 0$.

<u>Proof</u>: $\|\phi_\varepsilon - \phi_o\| \le \|L_o^\dagger\| \cdot \|f_o - f_\varepsilon\| + \|L_o^\dagger - F_\varepsilon^\dagger\| \cdot \|f_\varepsilon\|$. The first term goes to 0 because of (2.5), which also implies that $\|f_\varepsilon\|$ remains bounded.
By (2.10) and (2.11), $\|L_o^\dagger - F_\varepsilon^\dagger\| \le \|L_o^\dagger\| \cdot \sum_{i=1}^{\infty} \|M_\varepsilon\|^i = \|L_o^\dagger\| \cdot (1-\|M_\varepsilon\|)^{-1} \cdot \|M_\varepsilon\| \to 0$ as $\varepsilon \to 0$ because of (2.6) and (2.7). □

While ϕ_o is a solution of $L_o x = f_o$ (since $f_o \in R(L_o)$ by assumption), in general ϕ_ε will not solve $L_\varepsilon x = f_\varepsilon$. Instead, because of Lemma 2.3 and (1.6) we have $F_\varepsilon \phi_\varepsilon = (I-P)f_\varepsilon$, or equivalently,

(2.14) $$L_\varepsilon \phi_\varepsilon = (I-P)f_\varepsilon + PL_\varepsilon \phi_\varepsilon .$$

The next step reminds of Cesari's "alternative method" for nonlinear equations with non-invertible linear part. We split the equation $L_\varepsilon x=f_\varepsilon$ into two parts, one of which is (2.14). The solution of (2.14) has already the desired convergence properties. We have to make up for the fact that ϕ_ε is not a solution of $L_\varepsilon x = f_\varepsilon$ by adding a term, which is determined by a finite-dimensional equation:

Lemma 2.5: Let n be as in (2.1). For sufficiently small $\varepsilon > 0$, the equation

$$PL_\varepsilon (I-M_\varepsilon)^{-1}x = P(f_\varepsilon - L_\varepsilon \phi_\varepsilon) \tag{2.15}$$

has an n-dimensional linear manifold of solutions $x_\varepsilon \in N(L_0)$. An $x_\varepsilon \in N(L_0)$ solves (2.15) if and only if

$$\varphi_\varepsilon := \phi_\varepsilon + (I-M_\varepsilon)^{-1} x_\varepsilon \tag{2.16}$$

solves $L_\varepsilon x = f_\varepsilon$; all solutions of that equation can be written in the form (2.16), where $x_\varepsilon \in N(L_0)$ solves (2.15).

Proof: With the additional observation that $(I-M_\varepsilon)^{-1}(D) \subseteq D$ (see the proof of Lemma 2.3) and that therefore $\phi_\varepsilon \in D$, the proof is identical to the proof of Lemma 2.7 in [4]. □

Thus, in order to get convergence results for solutions of $L_\varepsilon x = f_\varepsilon$, it suffices to study the behaviour of solutions of (2.15) in $N(L_0)$ as $\varepsilon \to 0$. Since $\dim N(L_0) < \infty$, (2.15) is a finite-dimensional equation with a bounded linear operator that tends to 0 as $\varepsilon \to 0$; the same holds for the right-hand side of (2.15), since $P(f_\varepsilon - L_\varepsilon \phi_\varepsilon) = Pf_\varepsilon - PK_\varepsilon \phi_\varepsilon$, where K_ε is defined by (2.4). Because of the boundedness of K_ε, the assumption that $f_0 \in R(L_0)$, (2.3) and Lemma 2.4 we have $\lim_{\varepsilon \to 0} P(f_\varepsilon - L_\varepsilon \phi_\varepsilon) = 0$.

In this situation, the following result will be helpful:

Proposition 2.6: Let X_0 be a finite-dimensional Banach space. For all $\varepsilon > 0$, let $R_\varepsilon : X_0 \to X_0$ be a linear operator of the form

(2.17) $$R_\varepsilon = \varepsilon^i H + o(\varepsilon^i) ,$$

where $i \in \mathbb{N}$; $\dim N(R_\varepsilon)$ is assumed to be independent of ε for sufficiently small $\varepsilon > 0$. Assume that for those $\varepsilon > 0$ the linear operator H fulfills

(2.18) $$k := \dim N(H) \leq \dim N(R_\varepsilon) .$$

For all $\varepsilon > 0$, let $r_\varepsilon \in \mathcal{R}(R_\varepsilon)$ be of the form

(2.19) $$r_\varepsilon = \varepsilon^i r + o(\varepsilon^i)$$

with $r \in X_o$.
Then there exists a k-dimensional linear manifold in X_o with the property that all its elements are limits of solutions of

(2.20) $$R_\varepsilon x = r_\varepsilon$$

as $\varepsilon \to 0$. This manifold is the solution set of the (solvable) equation

(2.21) $$Hx = r,$$

no element outside this manifold is a limit of solutions of (2.20) under $\varepsilon \to 0$.

Proof: [4].

If we apply Proposition 2.6 to (2.15) in $X_o := N(L_o)$, we obtain the main result of this section:

Theorem 2.7: Let the general assumptions of this chapter about L_ε and f_ε $(\varepsilon \geq 0)$ be fulfilled. Assume that there exist a bounded linear operator $H: X \to X$ and an $i \in \mathbb{N}$ with

(2.22) $$K_\varepsilon = \varepsilon^i H + o(\varepsilon^i),$$

where K_ε is defined by (2.4), and that

$$\dim N(PH|_{N(L_o)}) \leq n, \tag{2.23}$$

where n is defined by (2.1).

Finally, assume that

$$Pf_\varepsilon = \varepsilon^i s + o(\varepsilon^i) \tag{2.24}$$

with $s \in X$.

Then there is an n-dimensional linear submanifold of the solution set of $L_o x = f_o$ all elements of which are limits of solutions of $L_\varepsilon x = f_\varepsilon$ as $\varepsilon \to 0$; this submanifold contains all such limits and is given by

$$\{x \in D \mid L_o x = f_o,\ PHx = s\}. \tag{2.25}$$

Proof: The proof is the same as the proof of Theorem 2.9 in [4] (with the misprints concerning the Landau symbols mentioned above corrected!), if the Lemmata used there are replaced by the ones given above. Finally, (2.25) follows from Remark 10 in [4] . □

Theorem 2.7 is the desired generalization of Theorem 2.9 of [4] to the case of not necessarily bounded operators L_ε and will be basic for the following treatment of the nonlinear problem.

3. The Nonlinear Problem

Let X be a real Banach space, $D \subseteq X$ dense. For all $\varepsilon \geq 0$, let $L_\varepsilon : D \to X$ be linear and closed, $N_\varepsilon : X \to X$ continuous and bounded, but not necessarily linear. For sake of simplicity we assume that for $\varepsilon > 0$, L_ε is bijective, so that $L_\varepsilon^{-1} : X \to D$ is bounded. In principle, everything we do below could also be done under assumptions on L_ε like the ones made in Section 2 ; we will comment on this later.

As in Section 2, we assume that K_ε (defined as in (2.4)) is continuous; again we identify K_ε with its unique bounded extension to X.
We want to investigate the behaviour of solutions of (1.1) as $\varepsilon \to 0$.
The assumptions on the limiting problem are the following:

$$\dim N(L_0) < \infty \tag{3.1}$$

$$R(L_0) \text{ closed, } X = N(L_0) \oplus R(L_0). \tag{3.2}$$

Let $P: X \to X$ denote the projector onto $N(L_0)$ induced by the decomposition in (3.2). By $L_0^\dagger$ we denote the generalized inverse of L_0 with respect to P, I-P (see also Lemma 2.1). Let $x_0 \in D$ be a solution of the limiting problem, i.e.,

$$L_0 x_0 = N_0(x_0). \tag{3.3}$$

There are numerous sufficient conditions known for such a solution to exist; we will give one below. We will only be able to handle our problem if N_ε is "not far" from $N(x_0)$ in the following sense:
We assume that for $\varepsilon > 0$ there are bounded, Lipschitz continuous functions $M_\varepsilon: X \to X$ such that for all $x \in X$ and $\varepsilon > 0$,

$$N_\varepsilon(x) = N_0(x_0) + c(\varepsilon) M_\varepsilon(x), \tag{3.4}$$

where $c: [0,+\infty] \to \mathbb{R}_0^+$. We assume there is a Lipschitz constant $\alpha > 0$ (independent of ε, as soon as $\varepsilon > 0$ is sufficiently small) for M_ε. Furthermore, we make the assumption that $\limsup_{\varepsilon \to 0} \|M_\varepsilon(x_0)\| < \infty$. This is reasonable, since in most interesting cases, $\varepsilon \to M_\varepsilon(x)$ will be continuous in ε for all $x \in X$ anyway.
Finally we assume that for all $\varepsilon > 0$, $L_\varepsilon^{-1} M_\varepsilon$ is compact. Since L_ε^{-1} is bounded, this is fulfilled e.g. if M_ε is compact; if on the other hand L_ε^{-1} is completely continuous (which will be frequently the case if L_ε is a differential operator), the assumption is also fulfilled.
Our last general assumption concerns the function c. We assume of course

that $\lim_{\varepsilon \to 0} c(\varepsilon) = 0$, but moreover that there is a $C > 0$ such that for sufficiently small $\varepsilon > 0$,

$$(3.5) \qquad c(\varepsilon) \leq C.\| L_\varepsilon^{-1} \|^{-1}$$

holds. This is not an unreasonable assumption, as the following argument should demostrate: If $\dim N(L_0) > 0$, then typically $\|L_\varepsilon^{-1}\| \to \infty$ as $\varepsilon \to 0$; if X is a finite-dimensional Euclidean space, $\|L_\varepsilon^{-1}\| \geq \|L_\varepsilon - L_0\|^{-1}$ (see [6]). Since (1.1) is equivalent to $x = L_\varepsilon^{-1} N_0(x_0) + c(\varepsilon) L_\varepsilon^{-1} M_\varepsilon(x)$, one cannot in general expect convergence if (roughly spoken) the growth in L_ε^{-1} is not compensated for by $c(\varepsilon)$. This can be seen already in the linear one-dimensional case.

Now we quote an existence result for nonlinear alternative problems obtained by L. Cesari and the author, that will be used below:

<u>Theorem 3.1</u> ([3]): Let L_0, P, L_0^+, and N_0 be as above and assume (3.1) and (3.2) are fulfilled and that $L_0^\dagger N_0$ is compact. Let $\bar{x}$ be an element in X. Assume that there are $R, r > 0$ such that

a) For $x \in X$ with $\|P(x-\bar{x})\| \leq R$ and $\|(I-P)(x-\bar{x})\| = r$ we have $\|N_0(x) - L_0\bar{x}\| \leq \|L_0^\dagger\|^{-1} r$.

b) For $x \in X$ with $\|P(x-\bar{x})\| = R$ and $\|(I-P)(x-\bar{x})\| \leq r$ we have $\|PN_0(x)\|^2 \geq \|P(x-\bar{x}-N_0(x))\|^2 - R^2$.

Then $L_0 x = N_0(x)$ has at least one solution x with $\|P(x-\bar{x})\| \leq R$ and $\|(I-P)(x-\bar{x})\| \leq r$, i.e., $\|x-\bar{x}\| \leq R+r$.

This result can be viewed as a result about the existence of an exact solution in a neighbourhood of an approximate solution $\bar{x}$; see [3] for a more detailed discussion, especially of the case $\bar{x} = 0$.

Our convergence result is now the following:

<u>Theorem 3.2</u>: Let $L_\varepsilon, N_\varepsilon, M_\varepsilon, c, C, P, \alpha, x_0$ be as above and let all general assumptions of this section be fulfilled. Assume that:

$$(3.6) \qquad \alpha.C < 1$$

and that there exist a bounded linear operator $H:X \to X$ and an $i \in \mathbb{N}$ with

$$(3.7) \qquad K_\varepsilon = \varepsilon^i H + o(\varepsilon^i),$$

where K_ε is defined as in (2.4), and that

$$(3.8) \qquad N(PH|_{N(L_0)}) = \{0\}.$$

Furthermore, assume that

$$(3.9) \qquad c(\varepsilon)PM_\varepsilon(x_0) = \varepsilon^i s + o(\varepsilon^i)$$

with $s \in X$ and that

$$(3.10) \qquad PHx_0 = s.$$

Then for sufficiently small $\varepsilon > 0$, $L_\varepsilon x = N_\varepsilon(x)$ has a solution x_ε and

$$(3.11) \qquad \lim_{\varepsilon\to 0} \|x_\varepsilon - x_0\| = 0.$$

Proof: For $\varepsilon \geq 0$, let

$$(3.12) \qquad f_\varepsilon := N_0(x_0) + c(\varepsilon)M_\varepsilon(x_0),$$

i.e., $f_\varepsilon = N_\varepsilon(x_0)$. We treat the linear problem

$$(3.13) \qquad L_\varepsilon x = f_\varepsilon$$

by Theorem 2.7. Since L_ε is assumed to be regular and since $L_0x_0 = N_0(x_0)$, $f_\varepsilon \in R(L_\varepsilon)$ for all $\varepsilon \geq 0$. Note that $Pf_\varepsilon = PN_0(x_0) + c(\varepsilon)PM_\varepsilon(x_0) = PL_0x_0 + c(\varepsilon)PM_\varepsilon(x_0) = c(\varepsilon)PM_\varepsilon(x_0)$. Thus, the assumptions of Theorem 2.7 are fulfilled with $n = 0$; because of (3.10), the manifold of "approximable solutions" in (2.25) consists of the single point x_0. Thus it follows from Theorem 2.7, that with

(3.14) $$\bar{x}_\varepsilon := L_\varepsilon^{-1} f_\varepsilon$$

we have

(3.15) $$\lim_{\varepsilon \to 0} \| \bar{x}_\varepsilon - x_0 \| = 0.$$

Now we apply Theorem 3.1 to verify the existence of a solution of the nonlinear problem $L_\varepsilon x = N_\varepsilon(x)$ "close enough" to the solution $\bar{x}_\varepsilon$ of the linear problem $L_\varepsilon x = N_\varepsilon(x_0)$. We apply Theorem 3.1 to the following situation:

$L_\varepsilon, N_\varepsilon$ and $\bar{x}_\varepsilon$ ($\varepsilon > 0$) replace L_0, N_0 and $\bar{x}$ there. The condition b of Theorem 3.1 is void; since $N(L_\varepsilon) = \{0\}$, condition a reads as follows: "For $x \in X$ with $\|x-\bar{x}_\varepsilon\| \leq r_\varepsilon$ we have $\|N_\varepsilon(x)-L_\varepsilon \bar{x}_\varepsilon\| \leq \|L_\varepsilon^{-1}\|^{-1} r_\varepsilon$".

As soon as we have established this, the existence of a solution x_ε of $L_\varepsilon x = N_\varepsilon(x)$ with $\|x_\varepsilon - \bar{x}_\varepsilon\| \leq r_\varepsilon$ follows for every $\varepsilon > 0$. If we succeed in fulfilling condition a with r_ε such that

(3.16) $$\lim_{\varepsilon \to 0} r_\varepsilon = 0,$$

then $\lim_{\varepsilon \to 0} \|x_\varepsilon - \bar{x}_\varepsilon\| = 0$, which implies together with (3.15) that (3.11) holds. Therefore, all that remains to be done is to find r_ε with (3.16) such that condition a in the form given above is fulfilled. Let for $\varepsilon > 0$

(3.17) $$r_\varepsilon := \frac{C\alpha}{1-C\alpha} \cdot \|\bar{x}_\varepsilon - x_0\|.$$

Because of (3.6) and (3.15), (3.16) holds. Let $\varepsilon > 0$, $x \in X$ be such that $\|x-\bar{x}_\varepsilon\| = r_\varepsilon$. Then $\|N_\varepsilon(x)-L_\varepsilon \bar{x}_\varepsilon\| = \|N_\varepsilon(x)-N_\varepsilon(x_0)\| = \|c(\varepsilon)(M_\varepsilon(x)-M_\varepsilon(x_0))\|$ holds by the definitions of $\bar{x}_\varepsilon$ and N_ε. Because of (3.5), for sufficiently small $\varepsilon > 0$ we have:

$\|c(\varepsilon)(M_\varepsilon(x)-M_\varepsilon(x_0))\| \leq C.\|L_\varepsilon^{-1}\|^{-1}.\alpha(\|x-\bar{x}_\varepsilon\|+\|\bar{x}_\varepsilon-x_0\|) =$

$= C.\ \|L_\varepsilon^{-1}\|^{-1}\alpha(r_\varepsilon+\|\bar{x}_\varepsilon-x_0\|) = \|\bar{x}_\varepsilon-x_0\|.\|L_\varepsilon^{-1}\|^{-1}.C\alpha.\ (\frac{C\alpha}{1-C\alpha}+1) =$

$= r_\varepsilon.\|L_\varepsilon^{-1}\|^{-1}$ by (3.17). Thus, $\|N_\varepsilon(x)-L_\varepsilon\bar{x}_\varepsilon\| \leq r_\varepsilon.\|L_\varepsilon^{-1}\|^{-1}$ holds, which completes the proof by the remarks of above.□

Remark 3.3: If (3.5) is replaced by the stronger condition that $c(\varepsilon) = o(\|L_\varepsilon^{-1}\|^{-1})$, then (3.6) is superfluous, the result of Theorem 3.2 holds with any $\alpha > 0$. It should be noted that the two conditions imposed on the rate of decay of $c(\varepsilon)$, namely (3.5) and (3.9), are compatible, since they both give bounds in the same direction. In spaces where $\|L_\varepsilon^{-1}\| \geq \frac{K}{\|L_\varepsilon-L_0\|}$ holds if $N(L_0) \neq \{0\}$ with a suitable $K > 0$ (like in finite-dimensional Euclidean spaces, where $K = 1$, see [6]), (3.5) "nearly implies" (3.9), if (3.7) is assumed, as the following argument shows: Because of (3.7), $\|L_\varepsilon-L_0\| = O(\varepsilon^i)$ and thus $\|L_\varepsilon^{-1}\|^{-1} = O(\varepsilon^i)$; therefore (3.5) implies $c(\varepsilon) = O(\varepsilon^i)$, so that at least the same asymptotic behaviour as required in (3.9) holds automatically. If again (3.5) is replaced by the stronger condition $c(\varepsilon) = o(\|L_\varepsilon^{-1}\|^{-1})$, then the same argument shows that in the spaces considered here, (3.9) is really implied by this stronger condition (with $s = 0$).

In Theorem 3.2, we needed a condition singling out a particular solution x_0 of $L_0x = N_0(x_0)$, namely (3.10). This condition further restricts the admissible nonlinearities. If $c(\varepsilon) = O(\varepsilon^i)$, which is implied by (3.9) if $\|PM_\varepsilon(x_0)\|$ is bounded away from 0, (3.10) is in fact necessary for the conclusion to hold:

Proposition 3.4: Let $c(\varepsilon) = O(\varepsilon^i)$ and the conclusion of Theorem 3.2 hold. Then (3.10) is implied by the remaining assumptions of Theorem 3.2.

Proof: For $\varepsilon \geq 0$, let x_ε be as in Theorem 3.2. For $\varepsilon > 0$, $L_\varepsilon x_\varepsilon = N_\varepsilon(x_\varepsilon)$. Because of (3.7) and (3.4), this implies together with (3.11), that $L_0x_\varepsilon+\varepsilon^i Hx_\varepsilon+o(\varepsilon^i) = N_0(x_0)+c(\varepsilon)M_\varepsilon(x_\varepsilon)$. By applying the projector P and multiplying with ε^{-i}, we obtain

(3.18) $$PHx_\varepsilon + o(1) = c(\varepsilon).\varepsilon^{-i} PM_\varepsilon(x_\varepsilon),$$

since $PN_o(x_o) = PL_o x_o = 0$. Now $\|PM_\varepsilon(x_\varepsilon) - PM_\varepsilon(x_o)\| \leq \|P\|.\alpha\|x_\varepsilon - x_o\| =$ $= o(1)$ because of (3.11). Combining this with (3.18), we obtain

(3.19) $$PHx_\varepsilon + o(1) = c(\varepsilon)\varepsilon^{-i} PM_\varepsilon(x_o) + o(1),$$

which implies together with (3.9)

(3.20) $$PHx_\varepsilon + o(1) = s + o(1).$$

If we take $\varepsilon \to 0$ in (3.20) and use (3.11) and the boundedness of H, (3.10) follows. □

Remark 3.5: In principle, our way of proof for Theorem 3.2 would be applicable also if L_ε were not bijective. The conditions about the linear operators in Theorem 3.2 could be modelled after the analogous conditions in Theorem 2.7. However, two steps in the proof of Theorem 3.2 which in the version given were fulfilled automatically would have to be added to the assumptions: First, if f_ε were defined as in (3.12), we would have to assume that $f_\varepsilon \in R(L_\varepsilon)$ for sufficiently small $\varepsilon > 0$. Instead of defining $\bar{x}_\varepsilon$ by (3.14), we would have to define it just as a solution of $L_\varepsilon x = N_\varepsilon(x_o)$ coming from Theorem 2.7. In the "nonlinear part" of the proof of Theorem 3.2, condition b in Theorem 3.1 would not be void any more; an assumption would have to be added to ensure that with a suitable $R_\varepsilon \to 0$, this condition b could be fulfilled;note that the projector in this condition (onto $N(L_\varepsilon)$) might depend on ε. Finally, L_ε^{-1} would have to be replaced by $L_\varepsilon^\dagger$ throughout.

To summarize, we think that our assumption that for $\varepsilon > 0$, L_ε is invertible, can be easily replaced by the assumptions on L_ε of Section 2, but at the expense of additional conditions concerning the nonlinearity coming from Theorem 3.1. It does not seem to be worth these technical difficulties to write down a result for non-invertible L_ε here. It should be stressed, however, that condition b in Theorema 3.1, from which these difficulties come, is a rather natural condition (see [3]).

We close by illustrating our results with a simple example concerning the existence of periodic solutions for a nonlinear differential equation depending on a parameter.

Example 3.6: In [3], a result about existence of 2π-periodic solutions of

$$(3.21) \qquad x'(s)+g(x(s)) = h(s) \qquad (s \in \mathbb{R})$$

was proven as illustration for the abstract result of that paper (i.e., our Theorem 3.1). To illustrate the results of this paper, we consider a solution x_0 of (3.21) and study its stability under perturbations of the equation as treated here; more precisely, we study the question of uniform convergence of 2π-periodic solutions of

$$(3.22) \qquad x'(s)+\varepsilon x(s)+g(x_0(s)) = \varepsilon\bar{g}(x(s))+h(s) \qquad (s \in \mathbb{R})$$

to x_0. Here, g and $\bar{g}$ are assumed to be bounded and continuous, h continuous and 2π-periodic. We assume that $\bar{g}$ is Lipschitz continuous with constant $\alpha < 1$. Let x_0 be a 2π-periodic solution of (3.21), to which $\bar{g}$ is assumed to be related in such a way that the mean-value is preserved, i.e.

$$(3.23) \qquad \int_0^{2\pi} \bar{g}(x_0(s))ds = \int_0^{2\pi} x_0(s)ds$$

holds. We claim that for sufficiently small $\varepsilon > 0$, (3.22) has a 2π-periodic solution x_ε, for which

$$(3.24) \qquad \lim_{\varepsilon\to R}\ \sup_{s\in\mathbb{R}} |x_\varepsilon(s)-x_0(s)| = 0$$

holds. We show this by verifying the assumptions of Theorem 3.2. Let X be the Banach space of 2π-periodic functions from $\mathbb{R}$ into itself with the supremum-norm, $D: = \{x \in X | x \text{ continuously differentiable}\}$, $L_\varepsilon: D \to X$ be defined by $L_\varepsilon x: = x' + \varepsilon x$ $(\varepsilon \geq 0)$. For all $\varepsilon \geq 0$, L_ε is closed, $N(L_o) = \{x \in X | x \text{ constant}\}$, $R(L_o) = \{x \in X | \int_o^{2\pi} x(s)ds = 0\}$. Thus, (3.1) and (3.2) hold. The projector induced by (3.2) is given by

$$(3.25) \qquad (Px)(t): = \frac{1}{2\pi} \int_o^{2\pi} x(s)ds.$$

By the variation of constants formula, it can be easily seen that for arbitrary $f \in X$ and $\varepsilon > 0$, $L_\varepsilon x = f$ has the unique 2π-periodic solution

$$(3.26) \qquad (L_\varepsilon^{-1} f)(t): = e^{-\varepsilon t} \left[\int_o^t e^{\varepsilon s} f(s)ds + (e^{2\pi\varepsilon} - 1)^{-1} \cdot \int_o^{2\pi} e^{\varepsilon s} f(s)ds\right].$$

Thus, L_ε^{-1} exists and is completely continuous. Furthermore, it can be seen from (3.26), that for all $\varepsilon > 0$,

$$(3.27) \qquad \varepsilon \geq \|L_\varepsilon^{-1}\|^{-1} \geq \varepsilon\, e^{-2\pi\varepsilon}.$$

Since $\alpha < 1$ and $\lim\limits_{\varepsilon \to 0} e^{2\pi\varepsilon} = 1$, for sufficiently small $\varepsilon > 0$ it follows from (3.27) that (3.5) holds with $c(\varepsilon) = \varepsilon$ and $C = (2\alpha)^{-1}$. Thus, (3.6) holds. The condition (3.7) holds with $i = 1$ and $H = I$, which immediately implies (3.8). The nonlinear operators N_o, M_ε are given by

$$(3.28) \qquad N_o(x)(t): = -g(x(t)) + h(t)$$

and

$$(3.29) \qquad M_\varepsilon(x)(t): = \bar{g}(x(t)).$$

All the qualitative assumptions made for N_o and M_ε are fulfilled, $M_\varepsilon: X \to X$ is Lipschitz continuous with constant α for all $\varepsilon > 0$. The expression on the left-hand side of (3.9) is given by

$$(3.30) \qquad c(\varepsilon)PM_{\varepsilon}(x_o) = \frac{\varepsilon}{2\pi}\int_o^{2\pi} \bar{g}(x_o(t)dt,$$

so that in the notation of (3.9), $s = \frac{1}{2\pi}\int_o^{2\pi} \bar{g}(x_o(t))dt$. Thus, (3.10) is exactly our assumption (3.23). We can now apply Theorem 3.2 to conclude that for sufficiently small $\varepsilon > 0$, (3.22) has a 2π-periodic solution x_{ε} for which (3.11) and thus (3.24) holds.

References

[1] L. CESARI: Functional analysis and periodic solutions of nonlinear differential equations, in: Contributions to Differential Equations 1, Wiley, New York 1963, 149-187.

[2] L. CESARI: Functional analysis, nonlinear differential equations, and the alternative method, in: Nonlinear Functional Analysis and Differential Equations (L. Cesari, R. Kannan, J. Schuur, eds.), Dekker, New York 1976, 1-197.

[3] L. CESARI, H.W. ENGL : Existence and uniqueness of solutions for nonlinear alternative problems in a Banach space, Czechoslovak Math. Jour. 31 (106) (1981), 670-678.

[4] H.W. ENGL, R. KRESS: A singular perturbation problem for linear operators with an application to electrostatic and magnetostatic boundary and transmission problems, Math. Meth. in the Appl. Sc. 3 (1981), 249-274.

[5] M.Z. NASHED, G.F. VOTRUBA: A unified operator theory of generalized inverses, in: Generalized Inverses and Applications (M.Z. Nashed, ed.), Academic Press, New York 1976, 1-109.

[6] M.Z. NASHED: Perturbations and approximations for generalized inverses and linear operator equations, same volume, 325-396.

[7] H. WACKER: A summary of the developments on imbedding methods, in: Continuation Methods (H. Wacker, ed.), Academic Press, New York 1978, 1-35.

ON A PROPERTY OF ORLICZ-SOBOLEV SPACES

J.P. Gossez

Département de Mathématique, Campus Plaine - C.P. 214
Université Libre de Bruxelles, 1050 - Bruxelles, Belgique

1. The following question has been studied in the last years by Brézis and Browder [2], [3], [4]. Let u be a function in the Sobolev space $W_o^{m,p}(\Omega)$ and let S be a distribution in $W^{-m,p'}(\Omega) \cap L^1_{loc}(\Omega)$. Under what conditions is the function $S(x)\,u(x)$ integrable on Ω and if so, does $\int_\Omega S(x)u(x)dx$ equal $\langle S,u\rangle$ (where $\langle , \rangle$ denotes the pairing in the duality between $W^{-mp'}(\Omega)$ and $W_o^{m,p}(\Omega)$)?

It is our purpose here to describe and slightly improve part of our recent work [7] dealing with the extension of one of the results of [2] to the framework of Sobolev spaces built from Orlicz spaces.

Applications of theorem 1 below are given in [7] to variational boundary value problems for second-order quasi linear elliptic equations of the form

$$\sum_{|\alpha|\leq 1} (-1)^{|\alpha|} D^\alpha A\,(x,u,\nabla u) + g(x,u) = f.$$

Here the first term of the left-hand side is supposed to give rise to a "good" operator within the class of Orlicz-Sobolev spaces while the second term g satisfies a sign condition $g(x,u)u \geq 0$ but has otherwise unrestricted growth with respect to u. For instance the problem

$$-\sum_{i=1}^{N} \frac{\partial}{\partial x_i}\cdot \left(\phi\left(\frac{\partial^u}{\partial x_i}\right)\right) + g(u) = f \quad \text{in } \Omega,$$

$$u = 0 \text{ ou } \partial\Omega$$

can be handled in this way, where ϕ is any continuous, odd, increasing function from $\mathbb{R}$ to $\mathbb{R}$ and g any continuous function from $\mathbb{R}$ to $\mathbb{R}$ satisfying $g(u)u \geq 0$ on $\mathbb{R}$.

2. Let Ω be an open (possibly unbounded) subset of $\mathbb{R}^N$ which satisfies the segment property. Let $W^1L_M(\Omega)$ and $W^1E_M(\Omega)$ be the Orlicz-Sobolev spaces of order 1 on Ω corresponding to a N-function M. Standard references about these spaces include [1], [9]. They will, as usual, be identified to subsapces of the product $\Pi_{|\alpha|\leq 1} L_M(\Omega) = \Pi L_M$. Denoting by $\bar{M}$ the N-function conjugate to M, we define $W_o^1 L_M(\Omega)$ (resp. $W_o^1 E_M(\Omega)$) as the $\sigma(\Pi L_M, \Pi E_{\bar{M}})$ (resp.norm) closure of $\mathcal{D}(\Omega)$ in $W^1L_M(\Omega)$. We also consider $W^{-1}L_{\bar{M}}(\Omega)$ (resp.$W^{-1}E_{\bar{M}}(\Omega)$), the space of distributions on Ω which can be written as sums of a function in $L_{\bar{M}}(\Omega)$ (resp. $E_{\bar{M}}(\Omega)$) and of first order derivatives of functions in $L_{\bar{M}}(\Omega)$ (resp. $E_{\bar{M}}(\Omega)$). Since Ω has the following property, $\mathcal{D}(\Omega)$ is $\sigma(\Pi L_M, \Pi L_{\bar{M}})$ dense in $W_o^1 L_M(\Omega)$ (cf. [5], theorem 1.3) and consequently, the value of $S \in W^{-1} L_{\bar{M}}(\Omega)$ at an element $u \in W_o^1 L_M(\Omega)$ is well-defined. It will be denoted by $<S,u>$.

THEOREM 1. Let $S \in W^{-1}L_{\bar{M}}(\Omega) \cap L^1_{loc}(\Omega)$ and $u \in W_o^1 L_M(\Omega)$. Suppose that for some $H \in L^1(\Omega)$, $S(x)\ u(x) \geq h(x)$ a.e. in Ω. Then $Su \in L^1(\Omega)$ and $<S,u> = \int_\Omega S(x)\ u(x)dx$.

When $M(t) = |t|^p$ with $1 < p < \infty$, this result has been obtained in [2]. It then holds without any restriction on Ω. See also [3] for related results and [4] for the higher-order case. The introduction here of the (mild) assumption that Ω has the segment property is somehow related to the fact that, contrary to the situation in the case of the usual Sobolev spaces, the topology $\sigma(\Pi L_M, \Pi E_{\bar{M}})$ used in the definition of the space $W_o^1 L_M(\Omega)$ does not allow to go the limit on u in the pairing $<S,u>$. The main step in the proof of theorem 1 is the following approximation lemma. Once this is obtained, the arguments can be easily adapted from [2].

LEMMA 2. Let $u \in W_o^1 L_M(\Omega)$. Then there exists a sequence u_n such that (i) $u_n \in W_o^1 L_M(\Omega) \cap L^\infty(\Omega)$, (ii) supp u_n is compact in Ω, (iii) $|u_n(x)| \leq |u(x)|$ a.e. in Ω, (iv) $u_n(x)\ u(x) \geq 0$ a.e. in Ω, (v) $D^\alpha u_n \to D^\alpha u$ a.e. in Ω for $|\alpha| \leq 1$, (vi) for some $\lambda > 0$,

$$\int_\Omega M(D^\alpha u - D^\alpha u_n)/\lambda) \to 0 \text{ for } |\alpha| \leq 1.$$

We note that the convergence in (vi) (the so-called modular convergence) is stronger than that obtained in the corresponding result of [7]. Moreover the proof which is sketched below is simpler than that given in [7].

PROOF OF LEMMA 2. The following property of Orlicz-Sobolev spaces of order one will be used: if $v,w \in W_o^1 L_M(\Omega)$ and $h = \min\{v,w\}$, then $h \in W_o^1 L_M(\Omega)$ and

$$\frac{\partial h}{\partial x_i} = \begin{cases} \dfrac{\partial v}{\partial x_i} & \text{a.e. in } \{x \in \Omega \ ; \ v(x) \leq w(x)\}, \\[2ex] \dfrac{\partial w}{\partial x_i} & \text{a.e. in } \{x \in \Omega \ ; \ v(x) > w(x)\} \end{cases}$$

(see [7]). Writing $u = u^+ - u^-$, we can assume without loss of generality in the proof of lemma 2 that $u \geq 0$ a.e. in Ω. By theorem 4 in [6], there exists a sequence $v_n \in \mathcal{D}(\Omega)$ such that for $|\alpha| \leq 1$ and some $\lambda > 0$, $\int_\Omega M((D^\alpha u - D^\alpha v_n)/\lambda)dx \to 0$ as $n \to \infty$, and by the construction in [6], v_n can be taken ≥ 0 in Ω. Passing to a subsequence if necessary, we can assume that for $|\alpha| \leq 1$, $D^\alpha v_n \to D^\alpha u$ a.e. in Ω.

Define

$$u_n = \min\{u,v_n\}.$$

Then properties (i)-(iv) clearly hold. Computing the derivatives of u_n, we obtain

$$\frac{\partial u_n}{\partial x_i} = \begin{cases} \dfrac{\partial u}{\partial x_i} & \text{a.e. in } \Omega_n' = \{x \in \Omega; \ u(x) \leq v_n(x)\} \\[2ex] \dfrac{\partial v_n}{\partial x_i} & \text{a.e. in } \Omega_n'' = \{x \in \Omega; \ u(x) > v_n(\Omega)\}. \end{cases}$$

Thus (v) holds. To verify (vi), we write

$$\int_{\Omega} M((\frac{\partial u}{\partial x_i} - \frac{\partial u_n}{\partial x_i})/\lambda) = \int_{\Omega''_n} M((\frac{\partial u}{\partial x_i} - \frac{\partial v}{\partial x_i})/\lambda)$$

$$\leq \int_{\Omega} M((\frac{\partial u}{\partial x_i} - \frac{\partial v}{\partial x_i})/\lambda)$$

where the right hand side goes to zero. Q.E.D.

It is not known whether the above results (as well as their applications to strongly nonlinear boundary value problems) are valid for higher order Orlicz-Sobolev spaces. A proof along the above lines would require the extension of the Hedberg's delicate truncation procedure [8] to the framework of Orlicz spaces.

References

[1] R. ADAMS, Sobolev spaces, Ac. Press., New York, 1975.

[2] H. BREZIS and F. BROWDER, Sur une propriété des espaces de Sobolev C.R.Ac. Sc. Paris, 287 (1978),113-115.

[3] H. BREZIS and F. BROWDER, A property of Sobolev spaces, Com. Part. Diff. Eq. 9 (1979),1077-1083.

|4| H. BREZIS and F. BROWDER, Some properties of higher order Sobolev spaces, J. Math. Pures Appl., 61 (1982), 245-259.

[5] J.P. GOSSEZ, Nonlinear elliptic boundary value problem for equations with rapidly (or slowly) increasing coefficients, Trans. Amer. Math. Soc., 190 (1974),163-205.

[6] J.P. GOSSEZ, Some approximation properties in Orlicz-Sobolev spaces, Studia Mat., 74(1982), 17-24.

[7] J.P. GOSSEZ, Strongly nonlinear elliptic problems in Orlicz-Sobolev spaces of order one, Cah. Cent. Et. Rech. Oper., 24(1982), 221-228.

[8] L. HEDBERG, Two approximation properties in function spaces, Ark. Mat., 16 (1978),51-81.

[9] A. KUFNER, O. JOHN and S. FUCIK, Function spaces, Academia, Praha, 1977.

ANOTHER APPROACH TO ELLIPTIC EIGENVALUE PROBLEMS WITH RESPECT TO INDEFINITE WEIGHT FUNCTIONS

P. Hess and S. Senn
Mathematics Institute, University of Zürich,
Rämistr. 74, 8001 Zürich, Switzerland

1. In [1] the authors investigate the linear eigenvalue problem

$$\text{(EVP)} \qquad \mathcal{L} u = \lambda m(x) u \quad \text{in } \Omega, \qquad \frac{\partial u}{\partial \upsilon} = 0 \quad \text{on } \partial\Omega$$

in the bounded domain $\Omega \subset \mathbb{R}^N$ ($N \geq 1$) having smooth boundary $\partial\Omega$.
Here $\mathcal{L}$:

$$\mathcal{L} u = - \sum_{j,k=1}^{N} a_{ik} \frac{\partial^2 u}{\partial x_j \partial x_k} + \sum_{j=1}^{N} a_j \frac{\partial u}{\partial x_j}$$

is a uniformly elliptic differential expression of second order having real-valued coefficient functions $a_{jk} = a_{kj}$, $a_j \in C^{\Theta}(\bar{\Omega})$ $(0<\Theta<1)$, and $m \in C(\bar{\Omega})$ a given real-valued weight function, $m \not\equiv 0$. Without loss of generality we may assume $|m|<1$ on $\bar{\Omega}$. Further υ is an outward pointing, nowhere tangent smooth vector field on $\partial\Omega$, and λ denotes the eigenvalue parameter. Of course $\lambda=0$ is eigenvalue, with associated eigenfunction $u=\mathbb{1}$. One is concerned with the existence of nontrivial eigenvalues having a positive eigenfunction. [1] gives a complete answer to this question by reduction to the results of Hess-Kato [2] referring to the case in which the operator L (induced by $\mathcal{L}$ and the boundary condition) is invertible. The purpose of this note is to present a new and somewhat more transparent proof which is based (among others) in the following two facts: (i) a simple inequality for second order differential operators (Lemma 1), (ii) an interesting recent abstract result of Kato [3] which asserts that the principal eigenvalue $\mu(\lambda)$ of $\mathcal{L} - \lambda m$ is a concave function of λ. No specific reference to the results of [2] is necessary; these could be proved directly in a similar (even simpler) way as outlined here. We note that our present proof is particularly easy in what concerns the assertion on algebraic simplicity of the principal eigenvalues (Theorem 2(a)).

2. We employ the same notations and definitions as in [1]. Hence E denotes the real Banach space $C(\bar{\Omega})$ with norm $\|\cdot\|_E$ and $L : E \supset D(L) \to E$ the differential operator induced in E by $\mathcal{L}$ and the Neumann boundary conditions. Recall that L is the restriction to E of the operator L' given in the space $L^p(\Omega)$, p>N fixed, by

$$D(L') := \{u \in W^{2,p}(\Omega) : \frac{\partial u}{\partial \upsilon} = 0 \text{ on } \partial\Omega\}$$

$$L'u := u \quad (u \in D(L')).$$

Let $X := D(L)$, provided with the graph norm $\|u\|_X := \|u\|_E + \|Lu\|_E$. Then X is compactly and densely embedded in E. Further let $M : E \to E$ denote the multiplication operator by the function m. We define $u \neq 0$ to be an eigenfunction of the (EVP) to the eigenvalue $\lambda \in \mathbb{R}$ provided u solves

(1) $$Lu = \lambda Mu.$$

Besides (1) we also look at the equation

$(\hat{1})$ $$\hat{L}u = \hat{\lambda}\hat{M}u$$

($\hat{\lambda} \in \mathbb{C}$) obtained from (1) by complexification. If m does not change sign in Ω, (1) has no nontrivial eigenvalue having a positive eigenfunction [1,p.460]. Therefore we assume in the following that m admits both positive and negative values in Ω.

Since $\mathbb{1} \in N(L)$, as a consequence of the Krein-Rutman theorem [4] the (Banach space) adjoint operator $L^* : E^* \supset D(L^*) \to E^*$ has one-dimensional nullspace $N(L^*)$ spanned by a positive functional ψ.

The main results of [1] are :

Theorem 1. The spectrum $\sigma(\hat{L},\hat{M})$ of $\hat{L}$ with respect to $\hat{M}$ consists of at most countably many eigenvalues $\hat{\lambda} \in \mathbb{C}$ having no finite accumulation point.

Theorem 2. Suppose $\langle\psi,m\rangle\sigma \neq 0$. Then (1) admits a unique eigenvalue $\lambda_1(m) \neq 0$ having a positive eigenfunction. More precisely, $\lambda_1(m) \gtrless 0$ provided

$\langle\psi,m\rangle \gneq 0$. *Further*

(a) 0 *and* $\lambda_1(m)$ *are M-simple eigenvalues of* L;

(b) *if* $\lambda_1(m)>0$ *and* $\hat{\lambda}\in\mathbb{C}$ *is eigenvalue with* $\operatorname{Re}\hat{\lambda}>0$, *then* $\operatorname{Re}\hat{\lambda}\geqq\lambda_1(m)$ (*similar assertion if* $\lambda_1(m)<0$).

We recall that $\mu\in\mathbb{R}$ is K-simple eigenvalue of T ($K,T\in B(X,E)$) provided

(i) $\dim N(T-\mu K) = \operatorname{codim} R(T-\mu K) = 1$,

(ii) if $N(T-\mu K) = \operatorname{span}[u]$, then $Ku \notin R(T-\mu K)$.

Theorem 3. *Suppose* $\langle\psi,m\rangle = 0$. *Then* 0 *is the only eigenvalue of* (1) *having a positive eigenfunction*.

For the inhomogeneous equation

$$(L-\lambda M)u = h, \quad h \in E \text{ given}, \tag{2}$$

we have

Proposition 4. *Suppose* $\langle\psi,m\rangle<0$, *and that* (2) *holds with* $0<\lambda<\lambda_1(m)$. *Then* $h>0$ *implies* $u>0$.

Proposition 5. (i) *Suppose* $\langle\psi,m\rangle< 0$, *and that* (2) *holds for some* $h>0$, *with either* $\lambda \leqq 0$ *or* $\lambda \geqq \lambda_1(m)$. *Then* $u \not> 0$.

(ii) *Suppose* $\langle\psi,m\rangle = 0$, $h>0$, *and let* (2) *hold* (*with* $\lambda \neq 0$). *Then* $u \not> 0$.

Propositions 4 and 5(i) are to be modified in an obvious way if $\langle\psi,m\rangle>0$. Applications of the above results to nonlinear eigenvalue problems, particularly to an equation arising in a selection-migration model in population genetics, are contained in [5].

3. We first give a new proof of Theorem 1, based on the useful inequality stated in Lemma 1. Recall that by [1, Lemmata 1 and 2] $\hat{L}-\hat{\lambda}\hat{M}$ is a Fredholm operator (in the space $\hat{E}$) of index 0, and that it suffices to find $\hat{\lambda}_0 \in \mathbb{C}$ such that $\hat{L}-\hat{\lambda}_0\hat{M}$ is injective.

Lemma 1. *For* $w \in D(L)$ *we have also* $w^2 \in D(L)$, *and* $L(w^2) < 2wLw$ *provided* $w \notin N(L)$.

Proof. Note that $w^2\in W^{2,p}(\Omega)$ for $p > N$ (since $W^{2,p}(\Omega)$ is a Banach algebra for such p), and that w^2 satisfies the boundary conditions. Since

$$L'(w^2) = 2wL'w - 2 \sum_{j,k=1}^{N} a_{jk} \frac{\partial w}{\partial x_j} \frac{\partial w}{\partial x_k} \in E,$$

we conclude that $w^2 \in D(L)$ and $L(w^2) < 2wLw$ if $w \notin N(L)$. □

Lemma 2. 0 is isolated eigenvalue of (1) in $\mathbb{R}$.

Proof. We distinguish between the two cases $\langle \psi,m \rangle \neq 0$ and $\langle \psi,m \rangle = 0$.
(i) Let $\langle \psi,m \rangle \neq 0$. Suppose $(\lambda_j) \subset \mathbb{R}$ is a sequence of eigenvalues $\lambda_j > 0$ with $\lambda_j \to 0$, and $u_j \in E$ associated eigenfunctions, $\|u_j\|_E = 1$. Since $Lu_j = \lambda_j Mu_j \to 0$ in E, we have $\|u_j\|_X \leqq$ const. and may pass to a subsequence converging in E : $u_{jk} \to u$. As $\|u\|_E = 1$ and (since L is a closed operator in E) $Lu = 0$, we may assume $u = \mathbb{1}$. Thus

$$\langle \psi, Mu_{jk} \rangle \to \langle \psi, M\mathbb{1} \rangle = \langle \psi, m \rangle \neq 0.$$

On the other hand,

$$0 = \langle L^{*}\psi, u_{jk} \rangle = \langle \psi, Lu_{jk} \rangle = \lambda_{jk} \langle \psi, Mu_{jk} \rangle$$

implies $\langle \psi, Mu_{jk} \rangle = 0 \quad \forall k$, a contradiction.

(ii) Let $\langle \psi,m \rangle = 0$. We employ the decomposition $E = N(L) \oplus R(L)$ and assume again $(\lambda_j) \subset \mathbb{R}$ is a sequence of eigenvalues $\lambda_j > 0$ with $\lambda_j > 0$. Let the associated eigenfunctions $u_j = \alpha_j \mathbb{1} + v_j$, $v_j \in R(L)$, be normalized by $\alpha_j \geqq 0$, $\|v_j\|_E = 1$. Clearly

$$Lv_j = \lambda_j M(\alpha_j \mathbb{1} + v_j).$$

Two cases might occur:
(ii_1) the sequence $(\lambda_j \alpha_j)$ is unbounded in $\mathbb{R}$. Set $\tilde{v}_j := (\lambda_j \alpha_j)^{-1} v_j$. Then (for a subsequence) $\tilde{v}_j \to 0$ and $L\tilde{v}_j = m + \lambda_j M\tilde{v}_j \to m$ in E, and hence $L0 = m \neq 0$, which is impossible. Therefore

(ii_2) the sequence $(\lambda_j \alpha_j)$ is bounded in $\mathbb{R}$. Passing to a subsequence (without changing notation) we have $\lambda_j \alpha_j \to \gamma \geqq 0$, and since

$$Lv_j = \lambda_j \alpha_j m + \lambda_j Mv_j \to \gamma m,$$

we conclude that $\|v_j\|_X \leq$ const. Thus (for a further subsequence) $v_j \to v$ in E, with $v \in R(L)$, $\|v\|_E = 1$, and $Lv = \gamma m$. We infer that $\gamma = 0$ is impossible. If $\gamma > 0$, Lemma 1 implies that

$$L(v^2) < 2vLv = 2\gamma vm,$$

whence

$$0 = \langle \psi, L(v^2) \rangle < 2\gamma \langle \psi, Mv \rangle \tag{3}$$

(note that if $a \in E$, $a>0$, then $\langle \psi, a \rangle > 0$ since $\psi = ((L+1)^{-1})^* \psi$ and hence $\langle \psi, a \rangle = \langle \psi, (L+1)^{-1} a \rangle > 0$, as $(L+1)^{-1} a \in \mathrm{Int}(P_E)$). On the other hand

$$0 = \langle \psi, Lv_j \rangle = \lambda_j \langle \psi, Mv_j \rangle$$

and therefore in the limit

$$\langle \psi, Mv \rangle = 0. \tag{4}$$

(3) and (4) are incompatible. □

4. We now prove Theorems 2 and 3. Starting point is the following important result of Kato [3, Th.6.1]: the spectral bound

$$\gamma(\lambda) := \mathrm{spb}(-L+\lambda M) := \sup\{\mathrm{Re}\,\gamma : \gamma \in \sigma(-L+\lambda M)\}$$

is a convex function of $\lambda \in \mathbb{R}$. Thus

$$\mu(\lambda) = \inf\{\mathrm{Re}\,\mu : \mu \in \sigma(L-\lambda M)\}$$

is <u>concave</u> in $\lambda \in \mathbb{R}$ ($\mu(\lambda) = -\gamma(\lambda)$). By the Krein-Rutman theorem and a result of Protter-Weinberger [6] we know further that $\mu(\lambda)$ is the principal eigenvalue of $L-\lambda M$; hence there exists $u = u(\lambda) \in X \subset E$, $u > 0$, such that

$$(L-\lambda M)u(\lambda) = \mu(\lambda)u(\lambda). \tag{5}$$

Note that $\lambda \in \mathbb{R}$ is eigenvalue of (1) having a positive eigenfunction if and only if $\mu(\lambda) = 0$. We have $\mu(0) = 0$ and set $u(0) = \mathbb{1}$.
In a first step we prove some differentiability properties of μ and u with respect to λ. Let $J : E \supset X \to E$ be the injection mapping of X into E. Obviously (5) can be rewritten as

$$(5') \qquad (L-\lambda M)u(\lambda) = \mu(\lambda)\, Ju(\lambda).$$

Lemma 3. $\mu(\lambda)$ is J-simple eigenvalue of $L-\lambda M$.

Proof. $L-\lambda M-\mu(\lambda)J$ is Fredholm operator of index 0, and (by Krein-Rutman) $N(L-\lambda M-\mu(\lambda)\,J) = \mathrm{span}[u(\lambda)]$ as well as $N((L-\lambda M-\mu(\lambda)J)^*) = \mathrm{span}\,[\chi]$ for some $\chi \in E^*$, $\chi > 0$. Suppose now $Ju(\lambda) = (L-\lambda M-\mu(\lambda)J)w$ for some w. Then

$$0 = \langle \chi, (L-\lambda M-\mu(\lambda)J)w \rangle = \langle \chi, Ju(\lambda) \rangle > 0$$

since $Ju(\lambda) \in \mathrm{Int}(P_E)$. This contradiction proves Lemma 3. □

Considering for the moment L,M and J as operators in B(X,E), [7, Lemma 1.3] is applicable and guarantees that $\mu(\lambda)$ is an analytic function of λ, and that we can choose $u(\lambda) > 0$ locally in such a way that $\lambda \to u(\lambda) \in X$ is also analytic. Differentiating (5') with respect to λ, we obtain

$$(6) \qquad (L-\lambda M)u'(\lambda) - Mu(\lambda) = \mu(\lambda)Ju'(\lambda) + \mu'(\lambda)Ju(\lambda).$$

4A. Suppose $\langle \psi, m \rangle < 0$ (trivial modifications give the proof in case $\langle \psi, m \rangle > 0$). Setting $\lambda = 0$ in (6) we get

$$Lu'(0) - M\mathbb{1} = \mu'(0)J\mathbb{1}$$

and consequently

$$-\langle \psi, M\mathbb{1} \rangle = \mu'(0)\langle \psi, J\mathbb{1} \rangle .$$

We infer that

(7) $$\mu'(0) > 0.$$

Lemma 4. $\lim_{\lambda\to\pm\infty} \mu(\lambda) = -\infty$.

Proof. For $\lambda \to -\infty$, this follows trivially by the concavity of $\mu(\lambda)$ since $\mu'(0) > 0$.

By [1, Lemma 6] there exist $\lambda_o > 0$ and $w_o \in E$, $w_o > 0$, such that $w_o \leqq \lambda_o K_{\lambda_o} w_o$ (where the positive operator $K_{\lambda_o} : E \to E$ is defined by $K_{\lambda_o} := (L+\lambda_o)^{-1}(M+1)$). Thus $\frac{1}{\lambda_o} \leqq \operatorname{spr} K_{\lambda_o}$. Since

$$\left(\frac{1}{\lambda_o} - K_{\lambda_o}\right)u(\lambda_o) = \frac{\mu(\lambda_o)}{\lambda_o}(L+\lambda_o)^{-1}u(\lambda_o) \text{ with } u(\lambda_o) > 0 \text{ and}$$

$(L+\lambda_o)^{-1}u(\lambda_o) > 0$, it follows that $\mu(\lambda_o) \leqq 0$ by a well-known result (e.g. [8, Th. 2.16 and Th. 7.9]). $\lim_{\lambda\to+\infty} \mu(\lambda) = -\infty$ is now again a consequence of $\mu'(0) > 0$ and the concavity of $\mu(\lambda)$. □

In view of (7), Lemma 4 and the concavity, $\mu(\lambda) = 0$ precisely for $\lambda=0$ and some value $\lambda_1 = \lambda_1(m) > 0$. Further $\mu(\lambda) > 0$ precisely for $0<\lambda<\lambda_1$, and $\mu'(\lambda_1) < 0$.

Lemma 5. 0 and λ_1 are M-simple eigenvalues of L.

Proof. (i) $\lambda=0$: clearly $N(L) = \operatorname{span}[\mathbb{1}]$. Suppose $m=M\mathbb{1} = Lw$ for some w. Then $0 = \langle\psi, Lw\rangle = \langle\psi, m\rangle < 0$, a contradiction.

(ii) $\lambda=\lambda_1$: equation (6) at $\lambda=\lambda_1$ gives

(8) $$(L-\lambda_1 M)u'(\lambda_1) - Mu(\lambda_1) = \mu'(\lambda_1)Ju(\lambda_1).$$

Note that $N(L-\lambda_1 M) = \operatorname{span}[u(\lambda_1)]$, and assume $Mu(\lambda_1) \in R(L-\lambda_1 M)$. Since $\mu'(\lambda_1) < 0$, (8) implies that $Ju(\lambda_1) \in R(L-\lambda_1 M)$. But this contradicts Lemma 3 which says that $0 = \mu(\lambda_1)$ is J-simple eigenvalue of $L-\lambda_1 M$. □

We have established Theorem 2 except for assertion (b), whose proof now follows.

Suppose $\hat\lambda \in \mathbb{C}$ is eigenvalue of $(\hat 1)$ with $\operatorname{Re}\hat\lambda > 0$ and associated eigenfunction u. [1, Lemma 7] implies that $|u| \leqq (\operatorname{Re}\hat\lambda)K_{\operatorname{Re}\hat\lambda}|u|$, and thus (by the same argument as in the proof of Lemma 4) $\mu(\operatorname{Re}\hat\lambda) \leqq 0$. Hence

Re $\hat{\lambda} \geqq \lambda_1$.

4B. Suppose $\langle\psi,m\rangle = 0$. Relation (6) at $\lambda=0$ gives $Lu'(0)-M\mathbb{1}=\mu'(0)J\mathbb{1}$ and thus

$$0 = \langle\psi, Lu'(0)-M\mathbb{1}\rangle = \mu'(0)\langle\psi, J\mathbb{1}\rangle.$$

Since $\langle\psi,1\rangle > 0$, we obtain $\mu'(0) = 0$.As μ is a concave function of λ and 0 is isolated eigenvalue of (1), we conclude that $\mu(\lambda) < 0$ $\forall\lambda\neq 0$. This proves Theorem 3.

5. Propositions 4 and 5 are immediate consequences of the following consideration. Let first $\lambda > 0$. Then (5) holds iff

$$(9) \qquad (I-\lambda K_\lambda)u(\lambda) = \mu(\lambda)\ (L+\lambda)^{-1}u(\lambda), \quad u(\lambda) > 0.$$

If $\mu(\lambda) > 0$, the right hand side of (9) lies in $\mathrm{Int}(P_E)$, and we infer that $1 > \mathrm{spr}(\lambda K_\lambda)$. Since (2) is equivalent to

$$(10) \qquad u = (I-\lambda K_\lambda)^{-1}(L+\lambda)^{-1}h,$$

$h > 0$ implies $u > 0$ (develop $(I-\lambda K_\lambda)^{-1}$ in a Neumann series).
If $\mu(\lambda) \leqq 0$, we have $1 \leqq \mathrm{spr}(\lambda K_\lambda)$, and

$$(10') \qquad (\frac{1}{\lambda} - K_\lambda)u = \frac{1}{\lambda}\ (L+\lambda)^{-1}h$$

with $h > 0$ has no positive solution u [8, Th. 2.16 and Th. 7.9].
The case $\lambda < 0$ is reduced to the above treatment by observing that (2) is equivalent to $Lu-(-\lambda)(-M)u = h$.

References

[1] S. SENN and P. HESS: On positive solutions of a linear elliptic eigenvalue problem with Neumann boundary conditions. Math.Ann.258 (1982), 459-470.

[2] P. HESS and T. KATO: On some linear and nonlinear eigenvalue prob-

lems with an indefinite weight function. Comm. P.D.E. 5 (1980), 999-1030.

[3] T. KATO: Superconvexity of the spectral radius, and convexity of the spectral bound and the type. Math. Z. 180(1982), 265-273.

[4] M.G. KREIN and M.A. RUTMAN: Linear operators leaving invariant a cone in a Banach space. A.M.S. Transl. 10(1962), 199-325.

[5] S. SENN: On a nonlinear elliptic eigenvalue problem with Neumann boundary condition, with an application to population genetics. Comm. P.D.E. 8 (1983), 1199-1228.

[6] M.H. PROTTER and H.F. WEINBERGER: On the spectrum of general second order operators. Bull. A.M.S. 72(1966), 251-255.

[7] M.G. CRANDALL and P.H. RABINOWITZ: Bifurcation, perturbation of simple eigenvalues and linearized stability. Arch. Rat. Mech. Anal. 52(1973), 161-180.

[8] M.A. KRASNOSELSKII: Positive solutions of operator equations. P. Noordhoff, Groningen 1964.

SOME RESULTS ON MINIMAL SURFACES WITH FREE BOUNDARIES

S. Hildebrandt

Mathematisches Institut der Universität Bonn

1. Introduction

In 1816, J.D. Gergonne [5] posed the following problem: "Couper un cube en deux parties, de telle manière que la section vienne se terminer aux diagonales inverses de deux faces opposées, et que l'aire de cette section, terminée à la surface du cube, soit un minimum. Donner, en outre, l'équation de la courbe suivant laquelle la surface coupante coupe chacune des autres faces de ce cube".

Already in the same year, Tédénat claimed that a solution will be furnished by some helicoid. This assertion was seriously doubted by Gergonne and, in fact, is not true.

It was but H.A. Schwarz [18] in 1872 who noted that a solution of Gergonne's problem must not only have mean curvature zero but has to meet the two faces of the cube, on which its boundary is not preassigned, under a right angle. More generally, he considered a surface M minimizing area among all surface bounded in part by given curves Γ, while the rest of its boundary lies on given surfaces S. By applying Gauss' formula for partial integration, he found that M has to intersect the "supporting" surfaces S orthogonally in a system of curves which one might call the "free trace" Σ of M on S.

Accordlingly, Schwarz formulated Gergonne's problem somewhat more generally as follows.

Determine the surface of mean curvature zero (i.e., minimal surfaces) bounded by two apposite faces S_1, S_2 of a cube and by a pair of straight arcs Γ_1, Γ_2 connecting four end points of these faces as depicted in Fig . 1, which intersect S_1 and S_2 orthogonally.
He found denumerably many simply connected minimal surfaces without singularities satisfying these boundary conditions. This might seem sur-

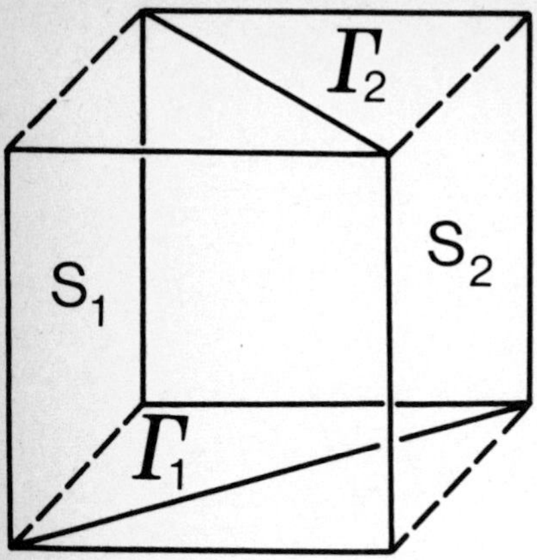

Fig. 1

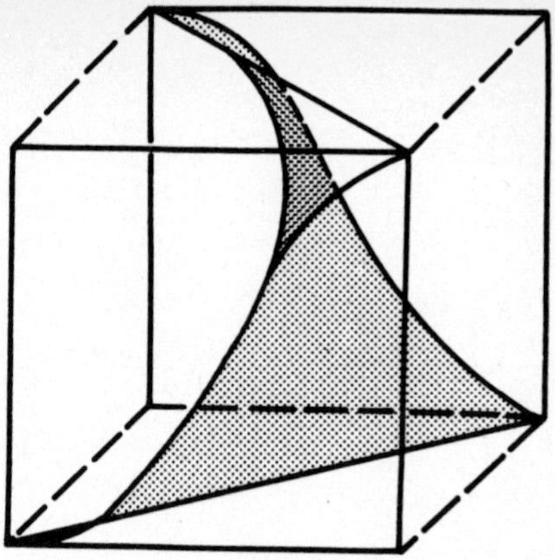

Fig. 2

prising since it is still not known as to whether there exist closed Jordan curves bounding more than finitely many simply connected minimal surfaces. On the other hand, one can easily find boundary configurations spanning more than denumerably many minimal surfaces which intersect the free part S orthogonally. For instance, a sphere, a cylinder, or a torus furnish an example, or, more generally, each surface S which is symmetrical with respect to a 1-parameter group of motions G and which bounds a minimal surface orthogonal to S but not invariant under G.

The Gergonne configuration from figure 1 does not possess such a symmetry so that Schwarz' result is a much more interesting one. A solution of Gergonne's problem is depicted in Fig. 2. Schwarz also described a further rigid configuration possessing infinitely many helicoids as solutions of the corresponding boundary value problem. In a cartesian system of coordinates x, y, z he considers a configuration $\langle \Gamma_1, \Gamma_2, S\rangle$ consisting of the cylinder surface

$$S = \{x^2 + y^2 = R^2 \ , \ |z| \leq \pi/4\}$$

and of the two straight arcs

$$\Gamma_1 = \{x = y \ , \ z = \pi/4\} \quad , \quad \Gamma_2 = \{x = -y \ , \ z = -\pi/4\} .$$

Then the parts of the helicoids

$$M_n^+ = \{\tan(2n+1)z = (-1)^n \frac{y}{x}\}$$

and

$$M_n^- = \{\tan(2n+1)z = (-1)^n \frac{x}{y}\} ,$$

$n=0, \pm 1, \pm 2, \ldots$, which are contained in the solid cylinder formed by the convex hull of S, are bounded by $<\Gamma_1, \Gamma_2, S>$ and meet S under a right angle. Their area $A_n^{\pm}$ is given by

$$A_n^{\pm} = \pi \int_0^R \sqrt{1+(2n+1)^2 r^2} \, dr$$

and is minimal for $n = 0$.

Since the fundamental investigations of Schwarz, a multitude of boundary value problems for minimal surfaces has been investigated, of which free boundary value problems play a prominent role. This does not only include the problem of finding minimal surfaces the boundary of which (or part of it) is left free on supporting manifolds, but also minimal surfaces with movable boundary curves of prescribed length, obstacle problems, and systems of minimal surfaces meeting along one or more curves whick are not preassigned.

We do not attempt to describe the variety of results on free boundary problems for minimal surfaces available in the literature. Instead we refer the reader to Chapter VI of J.C.C. Nitsche's lectures on minimal surfaces [17], to the papers [19], [20] and [10], [11], [12], [13] of Jean Taylor and of Hildebrandt-Nitsche.

In this survey, we restrict ourselves to the study of minimal surfaces having free or partially free boundaries on prescribed supporting sur-

faces.
Satisfactory results exist regarding the existence (1) of solution, and to the behavior of a solution surface near the fixed arcs of its boundary (2), while the behavior of a solution surface at its free boundary had only been studied for absolute minima of the area (3) but not for stationary solutions in general. Recent investigations by Grüter-Hildebrandt-Nitsche [8] and by Dziuk [3] have filled the gap, and the present survey will describe some of the results obtained in those papers. Moreover, we shall state an estimate for the length of the trace of a minimal surface on the free part of the boundary, derived by Hildebrandt-Nitsche [13].

2. Regularity at the boundary

The investigation of the boundary behavior of minimal surfaces in $\mathbb{R}^3$ with a free boundary can be reduced to the study of mixed boundary value problems for vector-valued functions

$$X(u,v) = (x^1(u,v),\ x^2(u,v),\ x^3(u,v))$$

satisfying a system of equations

$$(2.1)\qquad \Delta x^\ell + \Gamma^\ell_{ik}(X)\ \{x^i_u x^k_u + x^i_v x^k_v\} = 0$$

where $\Delta = \frac{\partial^2}{\partial u^2} + \frac{\partial^2}{\partial v^2}$ in the ordinary Laplacian, and Γ^ℓ_{ik} are the Christoffel symbols of second kind with respect to a symmetric positive definite matrix (g_{ik}). It turns out that the system (2.1) are the Euler equations of an integral of the form

$$(2.2)\qquad I_\Omega(X) = \int_\Omega f(u,v,X,X)\ du\ dv \quad , \Omega \subset \mathbb{R}^2 \quad ,$$

where $f(u,v,x,p)$ is a continuous function of its variables (u,v,x,p)

---

(1) Cf. [2], chapt. VI, pp. 199-223, and [17], Kap. VI, pp. 431-474.
(2) Cf. [17], Kap. V, pp. 281-348.
(3) Cf. [17], Kap. VI, pp. 447-474, and [10], [11], [12].

such that

$$(2.3) \qquad m_1|p|^2 - m_o \leqq f(u,v,x,p) \leqq m_2\,|p|^2 + m_o,\ 0 < m_1 \leqq m_2,\ m_o \geqq 0$$

and

$$f_{p^i_\alpha p^j_\beta}(u,v,x,p)\ \xi^i_\alpha\ \xi^j_\beta \geqq \lambda|\xi|^2 \quad , \qquad \lambda > 0\ ,$$

$$(2.4) \qquad |f_{p^i_\alpha p^j_\beta}| \leqq \mu$$

holds for all $(u,v) \in \Omega$, $x \in \mathbb{R}^3$, $p,\xi \in \mathbb{R}^6$ $(p = (p^i_\alpha)$, $\xi = (\xi^i_\alpha)$, $1 \leqq i \leqq 3$, $1 \leqq \alpha \leqq 2)$.

In fact, the integrand f associated with (2.1) is

$$(2.5) \qquad f(u,v,x,p) = g_{ik}(x)\{p^i_1 p^k_1 + p^i_2 p^k_2\} = g_{ik}(x) p^i_\alpha p^k_\alpha \quad .$$

Before we turn to the boundary regularity of stationary surfaces X of (2.2), we shall briefly review the situation with respect to interior regularity.

Morrey [16] has proved the following celebrated result:

<u>Suppose that</u> f <u>satisfies (2.3), and that</u> X $H^1_2(\Omega,\mathbb{R}^3)$ <u>fulfils</u>

$$I_\Omega(X) \leqq I_\Omega(x+\Phi) \quad \text{for all} \quad \Phi \in H^1_2(\Omega',\mathbb{R}^3)\ ,\ \Omega' \subset\subset \Omega.$$

<u>Then</u> X <u>is Hölder continuous in</u> Ω.

Starting from this point, one can prove higher regularity for every minimum X of I_Ω.

However, it is impossible to carry over Morrey's theorem to stationary points (2.1) in general since Frehse [4] has constructed an integrand

f satisfying (2.3) as well as the ellipticity condition (2.4), for which the integral (2.1) possesses a critical point X of class H_2^1. Thus it was un-clear whether "weak" minimal surfaces in a Riemmanian manifold are regular, i.e., are classical minimal surfaces.

However, the integral

$$(2.6)\qquad E_\Omega(X) = \int_\Omega g_{ik}(X)\,\{x_u^i x_u^k + x_v^i x_v^k\}\,du\,dv$$

is invariant with respect to conformal transformation of the independent variables. Thus it is well known that critical points of E_Ω with respect to boundary conditions of the Plateau type or to free boundary conditions are a. e. on Ω conformally parametrized, i.e., we have

$$(2.7)\qquad g_{ik}(X)\,x_u^i x_u^k = g_{ik}(X)\,x_v^i x_v^k\ ,\ g_{ik}(X)\,x_u^i x_v^k = 0 \quad \text{a. e. in } \Omega.$$

Thus weak minimal surfaces in a Riemannian manifold are weak solutions of (2.1), contained in H_2^1, and satisfying in addition (2.7).

Grüter has proved in his remarkable thesis [6,7] that one can derive from this fact the <u>interior regularity of each minimal surface in a Riemannian manifold</u>. The papers [8] and [3] employ Grüter's technique to tackle the regularity of minimal surfaces at a free boundary. The approach of [8] is somewhat more flexible and permits also the discussion of obstacle problems as considered in [10], [11], [12], while the method of [3] is based on the reflection principle introduced by Jäger [14] so that is cannot be applied to the obstacle problem. On the other hand, Dziuk's technique yields Hölder continuity of ∇X assuming only that the supporting surface S is of class $C^{1,\alpha}$.

In the following we shall restrict ourselves to the consideration of minimal surfaces with a partially free boundary. Since our approach will contain all the essential ideas, it can as well be applied to other free boundary value problems. Thus we consider the following type of boundary configurations in $\mathbb{R}^3$.

Let Γ be a regular arc in $\mathbb{R}^3$ having its two end points P_1 and P_2, $P_1 \neq P_2$, on a two-dimensional surface S of $\mathbb{R}^3$, but which has not other points in common with S.

We identify the two-dimensional Euclidean space $\mathbb{R}^2$ with $\mathbb{C}$, and write accordingly $w = (u,v) = u + iv$ for the points of $\mathbb{R}^2$. We shall choose the open semi-disc

$$B = \{w: |w| < 1, v > 0\}$$

as parameter domain of the surfaces $X = X(w)$ which will be considered.

Denote by C the closed circular arc $\{w: |w| = 1, v \geq 0\}$ and by I the open interval $\{w: |w| < 1, v = 0\}$, so that $\partial B = C \cup I$.

Then we introduce the class $\mathcal{C} = \mathcal{C}(\Gamma,S)$ of <u>admissible surfaces</u> $X = (x^1(w), x^2(w), x^3(w))$ as set of mappings $X \in H_2^1(B,\mathbb{R}^3)$ which are bounded by the configuration $\langle\Gamma,S\rangle$ in the following sense: for $X \in \mathcal{C}$ let X_C and X_I be the L_2-traces of X on C and I, correspondingly. Then X_C maps C continuously and in a weakly monotonic manner onto Γ such that $X_C(-1) = P_1$ and $X_C(1) = P_2$, while $X_I(w) \in S$ L^1-almost everywhere on I.

For $X \in H_2^1(B,\mathbb{R}^3)$ we introduce the Dirichlet integral by

$$D_B(X) := \int_B |\nabla X|^2 \, du \, dv$$

where $\nabla X = (X_u, X_v)$ is the weak gradient of X, and $|\nabla X| = \left(|X_u|^2 + |X_v|^2\right)^{1/2} = \left(x_u^i x_u^i + x_v^i x_v^i\right)^{1/2}$ denotes its Euclidean length.

As usual, a mapping $X: B \to \mathbb{R}^3$ is said to be a <u>minimal surface</u> (parametrized on the domain B) if it is real analytic, and if it satisfies Laplace's equation

$$\Delta X = 0$$

as well as the conformity relations

$$|X_u|^2 = |X_v|^2 \quad , \quad X_u \cdot X_v = 0$$

on B, and $X(w) \not\equiv$ const on B.

Furthermore we define an <u>admissible variation</u> of a surface $X \in \mathcal{L}$ as family $\{X_\varepsilon\}_{|\varepsilon|<\varepsilon_o}$, $\varepsilon_o > 0$, surfaces $X_\varepsilon \in \mathcal{L}$ with $X_o = X$ and such that $\eta = \lim_{\varepsilon\to 0} \frac{1}{\varepsilon} \{X_\varepsilon - X\}$ exists in $H_2^1(B, \mathbb{R}^3)$.

A surface $X \in \mathcal{L}$ is said to be <u>stationary in</u> $\mathcal{L}$ if

$$\lim_{\varepsilon\to 0} \frac{1}{\varepsilon} \left[D_B(X_\varepsilon) - D_B(X)\right] = 0$$

holds for all admissible variations of X.

It is known that a nonconstant surface $X \in \mathcal{L}$, which is stationary in $\mathcal{L}$, has to be a minimal surface (parametrized on B). We call it a minimal surface which is stationary in $\mathcal{L}$. It turns out that a stationary minimal surface intersects S orthogonally if it is of class C^1 at is free boundary.

In the following, we shall describe the behavior of minimal surface $X \in \mathcal{L}$, which are stationary in $\mathcal{L}$, at their <u>free boundary</u> I. For this purpose, we have at least to assume that S is a regular two-dimensional surface in $\mathbb{R}^3$, without self-intersections and without boundary, which is of class C^3. Moreover, we have to impose an assumption (V) which is a uniformity condition at infinity. This assumption will automatically be satisfied if S is also compact.

<u>Assumption (V)</u>

S is a two dimensional manifold of class C^2, imbedded into $\mathbb{R}^3$ and without boundary, for which there exist numbers $\rho_o > 0$, $K \geq 0$, and K_1, K_2 with $0 < K_1 \leq K_2$ such that the following holds:

For each $f \in S$, there exist a neighborhood U of f in $\mathbb{R}^3$ and a C^3-diffeomorphism h of $\mathbb{R}^3$ onto itself such that the inverse h^{-1} maps f onto 0, and U onto the open ball $\{y: |y| < \rho_o\}$ such that $S \cap U$ is mapped onto the set $\{y: |y| < \rho_o, y^3 = 0\}$ of the hyperplane $\{y^3 = 0\}$.

Moreover, if $g_{ik}(y) := h^l_{yi}(y) h^l_{yk}(y)$, (summation with respect to l from 1 to 3!), then we have

$$K_1|\xi|^2 \leqq g_{ik}(y)\xi^i\xi^k \leqq K_2|\xi|^2$$

for all $\xi \in \mathbb{R}^3$ and all $y \in \mathbb{R}^3$, and also $\left|\frac{\partial g_{ik}(y)}{\partial y^l}\right| \leqq K$

for all $y \in \mathbb{R}^3$ and all $i,k,l \in \{1,2,3\}$.

Finally, for every point $x^* \in \mathbb{R}^3$, there exists a point $f \in S$, such that $|x^*-f| = \operatorname{dist}(S,x^*)$ provided that $\operatorname{dist}(S,x^*) < \frac{1}{4}\rho_o\sqrt{K_2}$.

Now we can formulate the main result of this section.

Theorem 1. Suppose that $X: B \to \mathbb{R}^3$ is a minimal surface of class $\mathcal{C}(\Gamma,S)$ which is stationary in this class. Furthermore, let S be a supporting surface which satisfies assumption (V). Then X is of class $C^{2,\beta}(B \cup I, \mathbb{R}^3)$ for every $\beta \in (0,1)$.

Moreover, if $w_o \in I$ is a branch point of X on the free boundary, i.e. $X_w(w_o) = 0$, then there exist a vector $b = (b^1,b^2,b^3) \in \mathbb{C}^3$ with $b \neq 0$ and $b \cdot b = 0$, and an integer $\nu \geqq 1$, such that

$$X_w(w) = b\cdot(w-w_o) + o(|w-w_o|^\nu \quad \text{as} \quad w \to w_o$$

where $X_w = \frac{1}{2}(X_u - iX_v)$. Consequently, the surface normal

$$N(w) = \frac{X_u(w) \wedge X_v(w)}{|X_u(w) \wedge X_v(w)|}$$

tends to a limit vector as $w \to w_o$. That is, the tangent plane of X tends to a limiting position as w tends to a branch point on the free boundary. Moreover, the nonoriented tangent of the trace $\{X(w): w \in I\}$ of the minimal surface on S moves continuously through a boundary branch point. The oriented tangent is continuous at branch points w_o of even order ν, but, for branch points of odd order, the tangent direction jumps by 180° degrees.

Finally, X is of class $C^{s,\alpha}(B \cup I, \mathbb{R}^3)$ if also $S \in C^{s,\alpha}$, $s \geqq 2$, $0 < \alpha < 1$, and X is real analytic on $B \cup I$, if S is real analytic.

An interesting, non-planar, and not area minimizing but stationary minimal surface with boundary on S has been exhibited by H.A. Schwarz, Gesammelte Math. Abhandlungen I, pp. 149-150. We present the picture of this surface, due to Schwarz, in Fig. 3 :

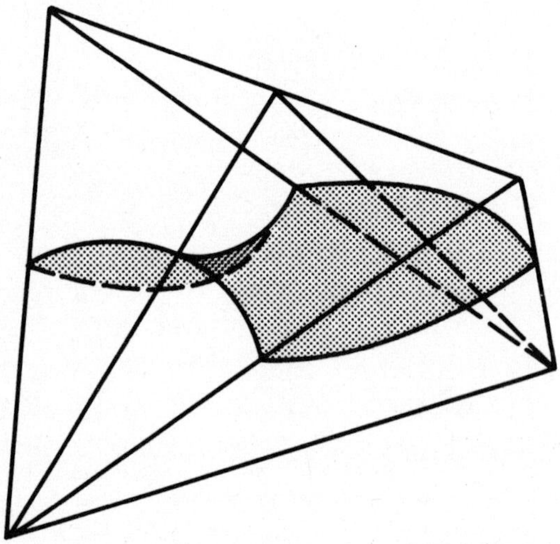

Fig. 3

We sketch the proof of Theorem 1, starting with the following fundamental observation.

Lemma 1. Let $X: B \to \mathbb{R}^3$ be a minimal surface. Then, for each open subset Ω of B and for every point $w^* \in \Omega$, we get

$$\limsup_{\sigma\to 0} \frac{1}{\sigma^2} \int_{\{w\in\Omega:\, |X(w)-X(w^*)|<\sigma\}} |\nabla X|^2 \, du\, dv \geq 2\pi .$$

The proof of this formula can easily be derived from the well known asymptotic expansion of minimal surfaces; cf. [17], §.361.

The next estimate is essentially due to Courant and Lebesgue (cf. [2], p. 102). We introduce the following notations:

Let $w = (u,v) = u + iv$ be a point of $\mathbb{R}^2 \cong \mathbb{C}$, and set $S_r(w_o) = \{w: |w-w_o| < r,\ v > 0\}$, $C_r(w_o) := \{w: |w-w_o| = r,\ v > 0\}$, $I := \{w: |w| < 1,\ v = 0\}$.

Lemma 2. For each $X \in C^1(B,\mathbb{R}^3)$, for every $w_o \in I$, and for each $R_o \in (0,1-|w_o|)$, there is a number $r \in [R_o/2, R_o]$ such that

$$\operatorname{osc}_{C_r(w_o)} X \leq \sqrt{\pi/\log 2} \left\{ \int_{S_{R_o}(w_o)} |\nabla X|^2 \, du\, dv \right\}^{1/2} .$$

The following estimate follows from a simple application of the triangle inequality:

Lemma 3. Let $w_o \in I$, $r \in (0,1-|w_o|)$, and suppose that X is a surface of class $C^1(B,\mathbb{R}^3)$. Assume also that, for some positive numbers α_1 and α_2,

$$\operatorname{osc}_{C_r(w_o)} X \leq \alpha_1$$

and

$$\sup_{w^*\in S_r(w_o)} \ \inf_{w\in C_r(w_o)} |X(w) - X(w^*)| \leq \alpha_2 .$$

Then

$$\operatorname{osc}_{S_r(w_o)} X \leq 2\alpha_1 + 2\alpha_2 .$$

The crucial estimate of our regularity result is contained in the following

Lemma 4. Let $w_o \in I$, *and suppose that* $X: B \to \mathbb{R}^3$ *is a minimal surface of class* $\tilde{\mathcal{L}} = \mathcal{L}(\Gamma, S)$ *which is stationary in* $\mathcal{L}$. *Assume also that the supporting surface* S *satisfies the assumption* (V) *with constants* ρ_o, K, K_1, K_2. *Then for* $\rho_1 := \rho_o \sqrt{K_2}$ *and for some number* K_3 *depending only on* ρ_o, K, K_1 *and on* K_2, *the following holds:*

If $w^* \quad S_r(w_o)$, $0 < r < 1 - |w_o|$, $0 < R < \rho_1$, *and if*

$$\inf_{w \in C_r(w_o)} |X(w) - X(w^*)| > R \quad ,$$

then

$$R \leqq K_3 \sqrt{e(w_o, r)}$$

where we have set

$$e = e(w_o, r) := \int_{S_r(w_o)} |\nabla X|^2 \, du \, dv \quad .$$

The proof of this lemma is rather complicated. We shall briefly indicate the main ideas of the proof at the end of this section and proceed presently with the *verification of Theorem 1*.

Choose an arbitrary point $w_o \in I$ and an arbitrary number R with $0 < R < \rho_1$. Since $D_B(X) < \infty$, we can find a number $R_o \in (0, 1 - |w_o|)$ such that

$$R > K_3 \sqrt{e(w_o, R_o)} \quad .$$

Then we infer from Lemma 4 that

$$\sup_{w^* \in S_r(w_o)} \inf_{w \in C_r(w_o)} |X(w) - x(w^*)| \leq R$$

for every $r \in (0,R_o)$.

Moreover, in virtue of Lemma 2, there exists a number $r \in \left[\frac{1}{2} R_o, R_o\right]$ such that

$$\operatorname{osc}_{C_r(w_o)} X \leqq K_4 \sqrt{e(w_o,R_o)} < (K_4/K_3) R$$

where $K_4 := \sqrt{\pi/\log 2}$.

On account of Lemma 3, we obtain that

$$\operatorname{osc}_{S_r(w_o)} X \leqq 2(1+K_4/K_3) R .$$

That is

$$\lim_{r\to 0} \operatorname{osc}_{S_r(w_o)} X = 0 .$$

Thus we have proved that X is continuous on $B \cup I$.

Next one proves by a "hole-filling" device that $X \in C^{0,\mu}(B \cup I, \mathbb{R}^3)$. This is by now more or less standard. For details, we refer the reader to [8], pp. 19-21.

From here on, well known techniques furnish the statement of Theorem 1; cf. [17], pp. 447-474 and p. 707 for references.

The <u>proof of Lemma 4</u> proceeds as follows.

Let $w^* \in S_r(w_o)$, $0 < R < \rho_1$, and set $x^* = X(w^*)$, $\delta(w^*) = \operatorname{dist}(S,x^*)$.

If $\delta(x^*) > 0$ we choose

$$\eta(w) = \begin{cases} 0 & \text{if } w \notin \overline{S_r(w_o)} \\ \lambda(\rho - |X(w)-x^*|)\{X(w)-x^*\} & \text{"} \quad w \in \overline{S_r(w_o)} \end{cases}$$

where $\lambda = \lambda_\varepsilon \in C^1(\mathbb{R}, \mathbb{R})$, $\lambda' \geqq 0$, $\lambda(t) = 0$ for $t \leqq 0$, and $\lambda(t) = 1$ if $t \geqq \tilde{\varepsilon}$.

It turns out that $X_\varepsilon = X + \varepsilon\eta$ is an admissible variation of X so that

$$0 = \lim_{\varepsilon\to 0} \frac{1}{\varepsilon} \{D_B(X_\varepsilon) - D_B(X)\} = \int_{S_r(w_o)} 2\nabla X\cdot\nabla\eta \, du\, dv .$$

Employing the conformality relations for X, letting $\tilde{\varepsilon}$ tend to zero, and taking Lemma 1 into account, we may infer that

$$2\pi \leqq R^{*-2} \int_{S_r(w_o)\cap K_{R^*}(x^*)} |\nabla X|^2 \, du\, dv$$

where we have set

$$R^* = \min\{\delta(x^*), d^2R\}, \quad K_\tau(X^*) = \{w \in B : |X(w)-x^*|<\tau\} .$$

Hence

$$(2.8) \qquad R \leqq \left\{ \frac{1}{2\pi d^4} \int_{S_r(w_o)} |\nabla X|^2 \, du\, dv \right\}^{1/2} \quad \text{if} \quad R^* = d^2R \ (\text{i.e. } d^2R=\delta(x^*)).$$

If $\delta(x^*) < d^2R$ (which will also include the case $\delta(x^*) = 0$) we have already found that

$$(2.9) \qquad 2\pi \leqq \delta(x^*)^{-2} \int_{S_r(w_o)\cap K_{\delta(x^*)}(x^*)} |\nabla X|^2 \, du\, dv .$$

Then there exists a point $f \in S$ such that

$$|f-x^*| = \delta(x^*) < d^2R \leqq \frac{1}{4} R .$$

We choose f as center of a new system of coordinates as indicated in assumption (V), with the defining diffeomorphism h, and we introduce $Y := h^{-1} \circ X$. Let $g_{ij}(Y) = h^\ell_{y^i}(Y)h^\ell_{y^j}(Y)$ be the components of the associated fundamental tensor, and set

$$\|Y(w)\|^2 := g_{ij}(Y(w)y^i(y)y^j(w) \quad .$$

If $d^{-1}\delta(x^*) < p < dR$, we define

$$\eta(w) = \begin{cases} 0 & \text{if } \begin{array}{l} w \notin B - \overline{S_r(w_o)} \\ w \in \overline{S_r(w)} \end{array} \\ \lambda(\rho \ \|Y(w)\|)Y(w) & \end{cases} .$$

For sufficiently small $|\varepsilon|$, the family of surfaces

$$X\ (w) = h(Y(w) + \varepsilon\eta(w))$$

forms an admissible variation of $X(w)$. Hence we infer from

$$\lim_{\varepsilon\to 0} \frac{1}{\varepsilon} \ \{D_B(X_\varepsilon) - D_B(X)\} = 0$$

that

$$\int_{S_r(w_o)} \{g_{ik}(Y)D_\alpha y^i D_\alpha \eta^k + \frac{1}{2} g_{ik,y^\ell}(Y)D_\alpha y^i D_\alpha y^k \eta^\ell\} \, du\, dv = 0$$

where $u^1 = u$, $u^2 = v$, $D_\alpha = \frac{\partial}{\partial u^\alpha}$, and $Y(w) = (y^1(w), y^2(w), y^3(w))$.

By a similar reasoning as before, we obtain that

$$\frac{1}{\delta^2(x^*)} \int_{S_r(w_o) \cap K_{2\delta(x^*)}(f)} |\nabla X|^2 \, du\, dv$$

(2.10)

$$\leqq \frac{C(R)}{d^4} \frac{1}{R^2} \int_{S_r(w_o)} |\nabla X|^2 \, du\, dv \quad .$$

By virtue of $K_{\delta(x^*)}(x^*) \subset K_{2\delta(x^*)}(f)$, we derive from (2.9) and (2.10) the inequality

$$(2.11) \qquad R^2 \leqq \frac{C(R)}{2\pi d^2} \int_{S_r(w_o)} |\nabla X|^2 \, du\, dv \quad \text{if} \quad 0 < \delta(x^*) < d^2 R.$$

In the case $\delta(x^*) = 0$ (i.e., $x^* = f$), we arrive at

$$(2.12) \qquad R^2 \leqq \frac{C(R)}{8\pi d^4} \int_{S_r(w_o)} |\nabla X|^2 \, du\, dv \quad \text{if} \quad \delta(x^*) = 0 \ .$$

From (2.8), (2.11), and (2.12), we infer that $R \leqq K_3 \sqrt{e}$ for some number K_3 as described in the assertion of Lemma 4, q.e.d..

3. <u>Geometric properties of the trace</u> .

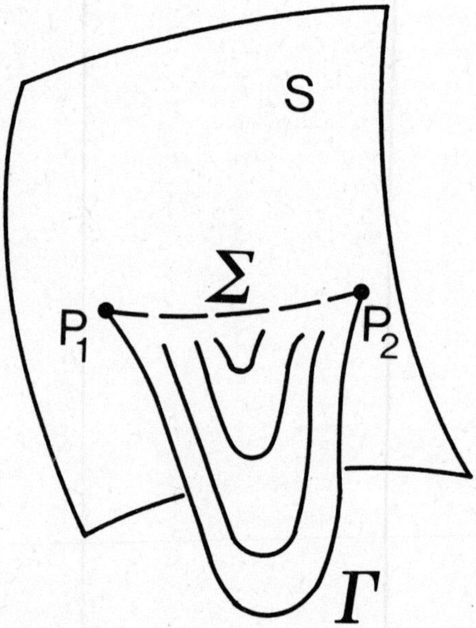

We consider in the following a boundary configuration $\langle\Gamma,S\rangle$ with the same properties as in section 2. Let $L(\Gamma) < \infty$ be the length of the arc Γ.

Moreover, let S be a regular orientable surface of class C^3 such that

(i) $\mathbb{R}^3 - S$ consists of two disjoint open sets;

(ii) for each point p of S, there are two spheres of radius R, one on each side of S, such that S has no points in common with the interior of these spheres.

Finally we assume that X is a minimal surface as in section 2, parametrized over the semidisc B, bounded by $\langle\Gamma,S\rangle$ and stationary for the Dirichlet integral in the class $\mathcal{L}(\Gamma,S)$.

Then the following holds (cf. [13]):

<u>Theorem 2</u>. <u>There exists a number</u> $c < 7/2$ <u>such that the length</u> $L(\Sigma) = \int_I |dX|$ <u>of the trace</u> $\Sigma = \{X(u,0): u \in I\}$ <u>can be estimated by</u>

$$L(\Sigma) \leqq L(\Gamma) + \frac{c}{R} D_B(X)$$

<u>provided that</u> X <u>possesses no branch points of odd order on</u> I.

<u>Remarks</u>.

1. In [13], also the following is proved:

(i) $X(u,v)$ has no branch points of odd order on I if it minimizes the Dirichlet integral in $\mathcal{L}(\Gamma,S)$.

(ii) X has no branch points at all <u>either</u> if S is real analytic, and $X(u,v)$ minimizes the Dirichlet integral in $\mathcal{L}(\Gamma,S)$, <u>or</u> if S is the boundary of an open, star-shaped and H-convex set (cf. [17], § 837), and Γ is contained in $\Omega \cup S$.

2. An immediate consequence of Theorem 2 is the following result:

$X(u,v)$ <u>is continuous on</u> $\bar{B} = B \cup C \cup I$ <u>if it does not possess any branch points of odd order on</u> I.

The proof of Theorem 2 follows from Gauss' formula

$$\int_B \nabla X \cdot \nabla \eta \, du\, dv = - \int_B \eta \cdot \Delta X \, du\, dv + \int_{\partial B} \frac{\partial X}{\partial \nu} \cdot \eta \, dH^1$$

by inserting the test function $\eta = \zeta(X)$, where ζ denotes a smooth vector field ζ on $\mathbb{R}^3$ which is on S of length one and orthogonal to S. Then we obtaine on I that

$$\frac{\partial X}{\partial u} \cdot \eta = |X_v| = |X_u|$$

whence

$$\int_I \frac{\partial X}{\partial u} \cdot \eta \, dH^1 = \int_I |X_u| \, du = L(\Sigma).$$

If we assume also $|\zeta| \leqq 1$ on $\mathbb{R}^3$, we get

$$\int_C \frac{\partial X}{\partial u} \cdot \eta \, dH^1 \leqq L(\Sigma)$$

and therefore

$$L(\Sigma) \leqq L(\Gamma) + \text{const } D_B(X).$$

A suitable choice of the vector field ζ leads to the more precise estimate stated in Theorem 2.

References

[1] R. COURANT: The existence of minimal surfaces of given topological structure under prescribed boundary conditions. Acta Math. 72, 51-98 (1940).

[2] R. COURANT: Dirichlet's principle, conformal mapping, and minimal surfaces. Interscience, New York 1950.

[3] G. DZIUK: Über die Stetigkeit teilweise freier Minimalflächen. Manuscripta math. 36, 241-251 (1981).

[4] J. FREHSE: Un problème variationel bidimensionel possédant des extremales bornées et dicontinues. C.R. Acad. Scienc. Paris.

[5] J.D. GERGONNE: Questions proposées. Ann. Math. Pure Appl. 7, 68, 99-100, 156 (1816).

[6] M. GRÜTER: Über die Regularität schwacher Lösungen des Systems $\Delta x = 2H(x)x_u \wedge x_v$. Dissertation, Düsseldorf 1979.

[7] M. GRÜTER: Regularity of weak H-surfaces. Journal für die Reine Angew. Math. 329, 1-15 (1981).

[8] M. GRÜTER, S. HILDEBRANDT, and J.C.C. NITSCHE: On the boundary behavior of minimal surfaces with a free boundary which are not minima of the area. Manuscripta math. 35, 387-410 (1981).

[9] S. HILDEBRANDT, and K.O. WIDMAN: Some regularity results for quasilinear elliptic systems of second order. Math. Z. 142, 67-86 (1975).

[10] S. HILDEBRANDT, and J.C.C. NITSCHE: Minimal surfaces with free boundaries. Acta Math. 143, 251-272 (1979).

[11] S. HILDEBRANDT, and J.C.C. NITSCHE: A uniqueness theorem for surfaces of least area with partially free boundaries on obstacles. To appear in Archive for Rat. Mech. Analysis.

[12] S. HILDEBRANDT, and J.C.C. NITSCHE: Optimal boundary regularity for minimal surfaces with a free boundary. Manuscripta math. 33, 357-364 (1981).

[13] S. HILDEBRANDT, and J.C.C. NITSCHE: Geometric properties of minimal surfaces with free boundaries. Preprint (1982).

[14] W. JÄGER: Behavior of minimal surfaces with free boundaries. Comm. Pure Appl. Math. 23, 803-818 (1970).

[15] H. LEWY: On minimal surfaces with partially free boundary. Comm. Pure Appl. Math. 4, 1-13 (1951).

[16] C.B. MORREY: Multiple integrals in the calculus of variations. Springer, Berlin-Heidelberg-New York 1966.

[17] J.C.C. NITSCHE: Vorlesungen über Minimalflächen. Springer, Berlin-Heidelberg-New York 1975.

[18] H.A. SCHWARZ: Fortgesetzte Untersuchungen über spezielle Minimalflächen. Monatsberichte der Königlichen Akad. Wiss. Berlin, 3-27 (1872). Gesammelte Math. Abhandlungen I, 126-148 (1890).

[19] J.E. TAYLOR: The structure of singularities in soap-bubble-like and soap-film-like minimal surfaces. Annals of Math. 103, 489-539 (1976).

[20] J.E. TAYLOR: Boundary regularity for various capillarity and boundary problems. Comm. P.D.E. 2, 323-357 (1977).

[21] K.O. WIDMAN: Hölder continuity of solutions of elliptic systems. Manuscripta math. 5, 299-308 (1971).

RELAXATION METHODS IN NONLINEAR PROBLEMS

R. Kannan

University of Texas at Arlington
Department of Mathematics
Arlington, Texas 76019

1. Introduction

The method of successive overrelaxation (SOR) for solving a system of linear equations Ax = b where A is a symmetric positive definite matrix is a well studied one [12,14]. In [11] the author studied this method for nonlinear problems and obtained sufficient conditions for the convergence of the method. We present here various results on the convergence of nonlinear successive overrelaxation which complement and generalize the literature in the linear theory. The detail of these results and various other ramifications may be seen in [1,2].

Our study of nonlinear successive overrelaxation was motivated by various questions. The first was the choice of the parameter which ensures convergence of the process. In particular in [5,6] the authors report divergence of nonlinear SOR for certain choices of the relaxation parameter when applied to the numerical solution of the minimal surface problem. The second question was the convergence of Newton (or related methods) at singular points [7,10]. We observe, by our method of proof, the convergence of nonlinear SOR for strictly convex functionals at singular points with a wider range for the parameter. A third problem is to try to utilize the results from the linear theory. We state here a theorem on convergence of nonlinear SOR where, by extensive utilization of the results from the linear theory, a different set of sufficient conditions is obtained.

In Section 2 of the paper we outline the linear theory for the sake of completeness. Section 3 briefly outlines some of the ideas in [5,6] on applying nonlinear SOR to the minimal surface problem. We present in Section 4 a survey of the results in [1,2] illustrating various features of nonlinear SOR and serving as a generalization of the results from the linear theory presented in Section 2.

2. Outline of linear SOR

In this section we outline for the sake of completeness some of the results from the theory of iterative methods for solving a system of linear equations that will be pertinent and related to the rest of the paper. For the sake of simplicity we first consider a system of three equations

$$
\begin{aligned}
a_{11}x_1 + a_{12}x_2 + a_{13}x_3 &= b_1, \\
a_{21}x_1 + a_{22}x_2 + a_{23}x_3 &= b_2, \\
a_{31}x_1 + a_{32}x_2 + a_{33}x_3 &= b_3,
\end{aligned}
\tag{2.1}
$$

where $a_{ii} \neq 0$, $i = 1,2,3$. These equations may then be also written as

$$
\begin{aligned}
x_1 &= \frac{1}{a_{11}} (b_1 - a_{12}x_2 - a_{13}x_3), \\
x_2 &= \frac{1}{a_{22}} (b_2 - a_{21}x_1 - a_{23}x_3), \\
x_3 &= \frac{1}{a_{33}} (b_3 - a_{31}x_1 - a_{32}x_2).
\end{aligned}
\tag{2.2}
$$

Let $x_i^{(n)}$ be the nth approximation to x_i. Assuming that $x_i^{(n)}$ is known, we can consider the following scheme to find $x_i^{(n+1)}$ which is known as the Gauss-Seidel method:

$$
\begin{aligned}
x_1^{(n+1)} &= \frac{1}{a_{11}} (b_1 - a_{12}x_2^{(n)} - a_{13}x_3^{(n)}), \\
x_2^{(n+1)} &= \frac{1}{a_{22}} (b_2 - a_{21}x_1^{(n+1)} - a_{23}x_3^{(n)}),
\end{aligned}
\tag{2.3}
$$

$$x_3^{(n+1)} = \frac{1}{a_{33}} (b_3 - a_{31}x_1^{(n+1)} - a_{32}x_2^{(n+1)}).$$

These equations may also be rewritten as

$$x_1^{(n+1)} = x_1^{(n)} + \frac{1}{a_{11}} |b_1 - a_{11}x_1^{(n)} - a_{12}x_2^{(n)} - a_{13}x_3^{(n)}|,$$

(2.4) $$x_2^{(n+1)} = x_2^{(n)} + \frac{1}{a_{22}} |b_2 - a_{21}x_1^{(n+1)} - a_{22}x_2^{(n)} - a_{23}x_3^{(n)}|,$$

$$x_3^{(n+1)} = x_3^{(n)} + \frac{1}{a_{33}} |b_3 - a_{31}x_1^{(n+1)} - a_{32}x_2^{(n+1)} - a_{33}x_3^{(n)}|.$$

and thus the terms in the square brackets in the right hand side of (2.4) may be viewed as the "improvement" on $x_i^{(n)}$ to obtain $x_i^{(n+1)}$. In the case when $A = (a_{ij})$ is a symmetric, positive definite matrix one can also associate with the system (2.1) a functional ϕ: $R^3 \to R^1$ defined by

(2.5) $$\phi(x_1,x_2,x_3) = \frac{1}{2} a_{11}x_1^2 + \frac{1}{2} a_{22}x_2^2 + \frac{1}{2} a_{33}x_3^2 + a_{12}x_1x_2 + a_{13}x_1x_3 + a_{23}x_2x_3 - b_1x_1 - b_2x_2 - b_3x_3.$$

It can be easily seen that ϕ is strictly convex and the system (2.1) exactly corresponds to the system of equations one would obtain in minimizing ϕ. It is this point of view that will be emphasized in this paper in the later sections. Further system (2.3) (or equivalently (2.4)) may also be viewed as follows: the first equation of (2.3) is the equation one obtains when minimizing $\phi(x_1,x_2^{(n)},x_3^{(n)})$ i.e., the coordinates x_2 and x_3 are kept fixed at $x_2^{(n)}$ and $x_3^{(n)}$ and one minimizes ϕ treated as a functional from R^1 to R^1 in the x_1 - direction. Similarly the second equation of (2.3) may be viewed as the gradient of ϕ in the x_2 - direction when x_1 is kept fixed at x_1^{n+1} and x_3 is fixed at x_3^n. We will utilize this approach to studying the Gauss-Seidel method for nonlinear problems later on.

Returning to system (2.4) and the remark following it, we note that if successive "corrections" are all one-signed it would be reasonable to expect that the convergence of $\{x^n\}$ would be accelerated if each equation of (2.4) was "corrected more". This situation of the successive corrections being one-signed arises naturally when one considers the finite-difference equations for a large class of elliptic problems. Thus we can now generate a modification of system (2.4) given by

$$x_1^{(n+1)} = x_1^{(n)} + \frac{\omega}{a_{11}} |b_1 - a_{11}x_1^{(n)} - a_{12}x_2^{(n)} - a_{13}x_3^{(n)}| ,$$

$$x_2^{(n+1)} = x_2^{(n)} + \frac{\omega}{a_{22}} |b_2 - a_{21}x_1^{(n+1)} - a_{22}x_2^{(n)} - a_{23}x_3^{(n)}| , \tag{2.6}$$

$$x_3^{(n+1)} = x_3^{(n)} + \frac{\omega}{a_{33}} |b_3 - a_{31}x_1^{(n+1)} - a_{32}x_2^{(n+1)} - a_{33}x_3^{(n)}| .$$

This iterative procedure for solving (2.1) is referred to as the method of successive overrelaxation and will be referred to as SOR for the rest of the discussions. As discussed after the Gauss-Seidel procedure, one could interpret the above equations in (2.6) in the case of a strictly convex functional as finding approximate minimum in successive coordinate directions instead of the exact minimum.

We now discuss the convergence of the Gauss-Seidel and SOR methods for a system of linear equations. Note that system (2.1) may be written as

$$Ax = b. \tag{2.7}$$

Let $A = D-L-U$ where $D = \begin{pmatrix} a_{11} & 0 & 0 \\ 0 & a_{22} & 0 \\ 0 & 0 & a_{33} \end{pmatrix}$, $L = \begin{pmatrix} 0 & 0 & 0 \\ -a_{21} & 0 & 0 \\ -a_{31} & -a_{32} & 0 \end{pmatrix}$

and

$$U = \begin{pmatrix} 0 & -a_{12} & -a_{13} \\ 0 & 0 & -a_{23} \\ 0 & 0 & 0 \end{pmatrix}.$$ Then (2.7) can be written as

$$(D-L-U)x = b. \tag{2.8}$$

It is now easy to see that the Gauss-Seidel procedure (2.3) can be written as

$$Dx^{(n+1)} = Lx^{(n+1)} + Ux^{(n)} + b$$

$$x^{n+1} = (D-L)^{-1}Ux^{(n)} + (D-L)^{-1}b. \tag{2.9}$$

The SOR procedure may be viewed as multiplying the correction in the Gauss-Seidel by a factor of ω. Thus in the Gauss-Seidel the correction is

$$\begin{aligned} x^{(n+1)}-x^{(n)} &= D^{-1}\,|Dx^{(n+1)}-Dx^{(n)}| \\ &= D^{-1}\,|Lx^{(n+1)}+Ux^{(n)}+b-Dx^{(n)}| \end{aligned}$$

Hence the SOR system may be written as

$$x^{(n+1)} - x^{(n)} = \omega D^{-1}|Lx^{(n+1)}+Ux^{(n)}+b-Dx^{(n)}|$$

or

$$\begin{aligned} (I-\omega D^{-1}L)x^{(n+1)} &= (1-\omega)x^{n} + \omega D^{-1}Ux^{(n)} + \omega D^{-1}b \\ &= \{(1-\omega)I+\omega D^{-1}U\}x^{(n)} + \omega D^{-1}b. \end{aligned}$$

Thus

$$x^{(n+1)} = (I-\omega D^{-1}L)^{-1}\{(1-\omega)I+\omega D^{-1}U\}x^{(n)} + (I-\omega D^{-1}L)^{-1}\omega D^{-1}b. \tag{2.10}$$

It is thus observed that both the Gauss-Seidel and SOR methods in the form expressed in (2.9) and (2.10) are of the type

$$x^{(n+1)} = Tx^{(n)} + c$$

and thus it suffices for the norm of T to be < 1 for some appropriate choice of the norm.

We now state without proof some results on convergence of SOR. For the details we refer to [12,14] . Note that Gauss-Seidel may be considered as a particular case of SOR when $\omega = 1$. We denote by L_ω the matrix

$$(I-\omega D^{-1}L)^{-1} \quad \{(1-\omega)I+\omega D^{-1}U\} \ .$$

Then we have:

<u>Theorem 2.1</u>. Let A be a symmetric positive definite matrix. Then, denoting by $\|A\|$ the spectral norm of A, for $0 < \omega < 2$,

i) $$\|L_\omega\|_{A^{\frac{1}{2}}} = \|A^{\frac{1}{2}}L_\omega A^{-\frac{1}{2}}\| < 1 \ ;$$

ii) the SOR process given by (2.6) or (2.10) converges to the solution of (2.1) for any starting point.

<u>Theorem 2.2.</u> Let A be a symmetric matrix with positive diagonal elements. Then the SOR process converges if and only if A is positive definite and $0 < \omega < 2$.

As remarked before, solving the system (2.1), when A is symmetric and positive definite, is equivalent to finding the point of minimum of the functional ϕ defined by (2.5) where ϕ is stricIty convex. By a simple application of Taylor's theorem, it is easy to see that:

<u>Theorem 2.3.</u> If A is a symmetric positive definite matrix, then, for $0 < \omega < 2$,

$$\phi(x^{(n+1)}) = \phi(x^{(n)}) - \frac{1}{2}\omega\,(2-\omega)(e^{(n)},e^{(n)})$$

where $e^{(n)} = x^{(n+1)} - x^{(n)}$. Further

$$\phi(x^{(n+1)}) < \phi(x^{(n)})$$

unless $x^{(n)}$ satisfies $Ax = b$.

The above theorem is critical in understanding our formulation of the nonlinear SOR results of the later sections. In fact, the above result states that for $0 < \omega < 2$, $\phi(x^{(n)})$ is decreasing and this is critical to how the relaxation parameter is to be chosen in the nonlinear case.

Two related questions arise. The first one is whether the order in which one chooses to minimize $\phi(x_1,x_2,x_3)$ coordinatewise or correspondingly the order in which the coordinate directions are chosen in the relaxation process is critical. In the linear theory that question has been answered in the affirmative in [9,14] and we will discuss this in more detail in the context of the nonlinear problem. The second question is whether the relaxation parameter ω could be varied. We now state the following:

<u>Theorem 2.4.</u> If A is a symmetric positive definite matrix, then the SOR process with relaxation parameters $\omega_1,\omega_2 \ldots$ converges provided at least one of the following holds:

a) for some $\varepsilon > 0$ we have

$$\varepsilon \leq \omega_i \leq 2 - \varepsilon;$$

b) $0 \leq \omega_i \leq 2$ for all i sufficiently large and $\Sigma\, \omega_i(2-\omega_i)$ diverges.

We finally recall from (2.10) that

$$(I-\omega D^{-1}L)x^{(n+1)} = |(1-\omega)I + \omega D^{-1}U|x^{(n)} + \omega D^{-1}b.$$

This can be further rewritten as

$$Bx^{(n+1)} + Cx^{(n)} = b \tag{2.11}$$

where $B = \frac{1}{\omega} D - L$ and $C = (1 - \frac{1}{\omega})D - U$.

One can then establish the convergence of the SOR process by showing that the spectral radius of $B^{-1}C$ is less than one. We now present a very interesting approach from [3,4] to generating (2.11) as an iterative process from the original problem $Ax = Bx + Cx = b$. This approach is then utilized in [3,4] to compare the convergence of various iterative methods. However we present only the generation of (2.11) in view of its possible extension to nonlinear problems and relation with homotopy approaches to studying global convergence for nonlinear problems. Thus let $x(\lambda)$ be a solution of

$$|B + \lambda C| x\ (\lambda) = b + (1-\lambda)e \tag{2.12}$$

where $\lambda \in C^1$ and e is an arbitrary vector. Further let the spectral radius of $B^{-1}C$ be less than or equal to $\gamma (<1)$. Then expanding $x(\lambda)$ we have

$$x(\lambda) = x(0) + x'(0)\lambda + x''(0)\lambda^2/2 + \ldots . \tag{2.13}$$

Note that $\det(B+\lambda C) = 0$ implies $|\lambda| \geq \frac{1}{\gamma}$. Hence from (2.12) and (2.13), setting $\lambda = 1$, we have

$$(B + C)x(1) = b \text{ i.e., } Ax(1) = b$$

and

$$x(1) = x(0) + x'(0) + x''(0)\lambda^2/2 + \ldots .$$

Let $X^0 = x(0)$, $\chi^1 = x(0) + x'(0)$, $X^2 = x(0) + x'(0) + \frac{x''(0)}{2}$ and so on.

Then we have

$$Bx(0) = b + e$$

(2.14) or $$x(0) = B^{-1}b + B^{-1}e.$$

Further, from (2.12),

(2.15) $$(B+\lambda C)x'(\lambda) + Cx(\lambda) = -e$$

or

(2.16) $$Bx'(0) + Cx(0) = -e.$$

From (2.14) and (2.15) we have

$$BX^1 + CX^0 = B(x'(0) + x(0)) + Cx(0) = b$$

or

$$X^1 = B^{-1}b - B^{-1}CX.$$

Proceeding similarly from (2.15) we have

$$(B + \lambda C)x''(\lambda) + 2C\ x'(\lambda) = 0$$

and

$$BX^1 + CX^1 = b.$$

In general, by induction, we have

$$BX^k + CX^{k-1} = b.$$

We note that X^k are the partial sums of $x(1) = x(0)+x'(0)+x''(0)\lambda^2/2+\ldots$ and thus are approximations to $x(1)$ which is the solution of $Ax=(B+C)x=b$.

3. Numerical solution of extrema of functionals

We discuss in this section two problems from the theory of extrema of functionals which have been studied in [5,6]. In these papers the authors have considered finite difference analogues and applied the SOR process discussed in the previous section to the finite difference problem.

Thus in [5] the author considers the problem of finding a continuously differentiable function $y(x)$ defined on $a \leq x \leq b$ satisfying $y(a) = \alpha$, $y(b) = \beta$ and minimizing the functional

$$J = \int_a^b F(x,y,y')dx .$$

The interval $[a,b]$ is divided into n parts by means of the points $x_0 = a, x_1, \ldots, x_{n-1}, x_n = b$ with $0 < h_i = x_i - x_{i-1}$. Setting $y_i = y(x_i)$, $i = 0,\ldots n$, we approximate the functional J by

$$J_n = \sum_{i=1}^{n} h_i F\left(x_{i-1}, y_{i-1}, \frac{y_i - y_{i-1}}{h_i}\right) .$$

Since $y_0 = \alpha$ and $y_n = \beta$, J_n is thus a function of $y_1,\ldots,y_{n-1}$. And thus to find an extremum of J_n one needs to solve the equations $\frac{\partial J_n}{\partial y_i} = 0$, $i = 1,\ldots,n-1$. As an example the author considers the problem where $a = 0$, $b = 1$, $y(0) = 1$, $y(1) = \cosh 1$ and

$$J = 2 \int_0^1 y\sqrt{1+(y')^2}\, dx .$$

We now recall that the SOR process applied to solve (2.1) or equivalently to minimizing the quadratic functional ϕ given by (2.5) may also be written as

$$x_1^{n+1} = x_1^n - \omega \frac{\partial \phi(x_1^n, x_2^n, x_3^n)}{\partial x_1} \Bigg/ \frac{\partial^2 \phi(x_1^n, x_2^n, x_3^n)}{\partial x_1^2} ,$$

$$x_2^{n+1} = x_2^n - \omega \frac{\partial\phi(x_1^{n+1},x_2^n,x_3^n)}{\partial x_2} \Bigg/ \frac{\partial^2\phi(x_1^{n+1},x_2^n,x_3^n)}{\partial x_2^2} ,$$

$$x_3^{n+1} = x_3^n - \omega \frac{\partial\phi(x_1^{n+1},x_2^{n+1},x_3^n)}{\partial x_3} \Bigg/ \frac{\partial^2\phi(x_1^{n+1},x_2^{n+1},x_3^n)}{\partial x_3^2} .$$

This process is now applied to study the functional $J_n : R^n \to R$ arising from the finite difference analogue of J. The author also studies the corresponding higher dimensional problem. Then let S be the square with vertices (0,0), (1,0), (1,1) and (0,1) and R_1 be its interior. Let $\phi(x,y) = (\cosh^2 x - y^2)^{\frac{1}{2}}$ for $(x,y) \in S$. The problem of finding a function $u(x,y)$, which is defined and continuous on $R_1 \in S$, is continuously differentiable on R_1, identical with ϕ on S and whose area over $R_1 \subset S$ is minimal, is equivalent to finding $u(x,y)$ with the above mentioned properties which minimizes

$$\iint (1+u_x^2+u_y^2)^{\frac{1}{2}} \, dA .$$

By using a difference approximation for u_x, u_y as before the author reduces it to minimization of a functional over R^N and applies the SOR process. Among the various test cases run by the author it is reported that when $h_{xi} \equiv h_{yi} \equiv .02$ and $\omega = 1.6$, the SOR process diverged.

The above study of the numerical solution of the minimal surface equation was done with the help of a higher order difference scheme in [6]. The Euler equation corresponding to minimizing the surface area

$$\iint (1+u_x^2+u_y^2)^{\frac{1}{2}} \, dx \, dy$$

is given by

$$\nabla\cdot(\gamma(|\nabla u|^2)\nabla u) = 0$$

where $\gamma(|\nabla u|^2) = (1+|\nabla u|^2)^{-\frac{1}{2}}$. For a square mesh of width $h = \frac{1}{N}$ the difference equation at an interior point is chosen as

$$f_{ij} = \gamma_{\bar{i}\bar{j}}(2u_{ij}-u_{i-1,j}-u_{i,j-1}) + \gamma_{\bar{i}+1,\bar{j}}(2u_{ij}-u_{i+1,j}-u_{i,j-1})$$

$$+ \gamma_{\bar{i},\bar{j}+1}(2u_{ij}-u_{i-1,j}-u_{i,j+1}) + \gamma_{\bar{i}+1,\bar{j}+1}(2u_{ij}-u_{i+1,j}-u_{i,j+1})$$

$$= 0, \quad 1 \leq i \leq N-1, \quad 1 \leq j \leq N-1$$

where $\gamma_{\bar{i}\bar{j}} = \gamma(|\nabla u|^2_{\bar{i}\bar{j}})$ denotes γ for the cell with center $(i-\frac{1}{2}, j-\frac{1}{2})$ and

$$|\nabla u|^2_{ij} = \frac{1}{2h^2}\Big|(u_{ij}-u_{i-1,j})^2 + (u_{ij}-u_{i,j-1})^2 + (u_{i,j-1}-u_{i-1,j-1})^2$$

$$+ (u_{i-1,j}-u_{i-1,j-1})^2\Big| .$$

The author [6] then applies the SOR process to the finite difference equations and once again among the many experimental results discussed, it is noted that when $h = \frac{1}{40}$ and Γ, the curve through which the surface of minimal area is sought is given by

$$x = 0, \; z = 0$$

$$x = 2, \; z = 0, \; 0 \leq y \leq 1$$

$$y = 1, \; z = 0$$

$$y = 0, \; z = 5 \lim(\pi x/2), \; 0 \leq x \leq 2,$$

the first 20 iterations of the SOR process were performed with $\omega = 1.9$ in order to prevent divergence.

Thus in both of the above examples the problem of safe ω to start the iteration is raised. There are several related questions e.g. optimal ω, varying ω at each stage and inexact SOR i.e., could one allow for errors in the SOR process. These questions lead us to study the convergence of the nonlinear SOR process.

4. Nonlinear SOR

In this section we present some of the results that have been obtained in [1,2] regarding various aspects of the convergence of nonlinear SOR. As mentioned in Section 1, the first part of the results are patterned and motivated after the results in [11].

We first make the following assumptions for the rest of this paper:

i) D is a convex domain in R^n;

ii) $\phi: D \to R$ is a strictly convex functional i.e., for all x,y in D and for every $\lambda \in (0,1)$,

$$\lambda\phi(x) + (1-\lambda)\phi(y) - \phi(\lambda x+(1-\lambda)y) \geq 0$$

with equality holding only when $x = y$.

iii) ϕ is twice continuously differentiable on D.

In addition to these assumptions on ϕ we make the following two important hypotheses:

iv) there exists $\gamma \in R$ such that $S_\gamma = \{x \in D : \phi(x) \leq \gamma\}$ is nonempty and compact;

v) $F_{ii}(y) \neq 0$ for $i = 1,\ldots,n$ and $y \in S_\gamma$, unless y is the point at which ϕ attains its minimum ($|F_{ij}(y)|$, $(i=1,\ldots,n;\ j=1,\ldots n)$ denotes the Hessian matrix of ϕ at y).

From ii) and iii) it is known that the Hessian matrix of ϕ is positive semidefinite. Also the sets S_γ are convex for all $\gamma \in R$. Further iii) implies that ϕ attains its minimum at some point $x^* \in S$. Conversely, if i), ii), and iii) are satisfied and ϕ attains its minimum at some point $x^* \in$ D then iv) is nontrivially satisfied i.e., there exists $\gamma \in R$ such that $\gamma > \phi(x^*)$ and S_γ is compact. By ii) the minimum point x^* is unique. Finally a point $x^* \in D$ is the minimum point of ϕ if and only if $\operatorname{grad} \phi(x^*) = 0$.

We first state and prove the nonlinear analogue of the Gauss-Seidel theorem for positive definite matrices: we refer to [13] also in this context.

Theorem 4.1. From any $x^0 = (x_1^0, x_2^0, \ldots, x_n^0)$ in S_γ we generate a sequence $\{x^k\}$ as follows:

$$x_j^{k+1} = x_j^k,\ j \neq i_k$$

(4.1) and

$$x_{i_k}^{k+1} \text{ is the solution of } F_{i_k}(x_1^k, x_2^k, \ldots, x_{i_{k-1}}^k, x_{i_k}, x_{i_{k+1}}^k, \ldots, x^n) = 0$$

where F_i denotes the component of grad ϕ in the ith coordinate direction and i_k is any one of the integers $1,2,\ldots,n$. Such a sequence $\{x^k\}$ is uniquely defined and converges to x^*, the unique global minimum of ϕ, provided that in the above iterative process every coordinate direction i is chosen an infinite number of times.

Proof. By the hypotheses on ϕ, for a given $x^k \in S_\gamma$ and $i_k \in (1,2,\ldots,n)$ there exists a unique point x^{k+1} satisfying (4.1). Also $x^{k+1} \in S_\gamma$ and hence $\{x^k\}$ is uniquely defined for a given $x^0 \in S_\gamma$ and sequence $\{i_k\}$. From the hypotheses it follows that $\phi(x^k) \geq \phi(x^{k+1}) \geq \phi(x^*)$ for all $k = 0,1,2,\ldots,$, and thus the subsequence $\{\phi(x^k)\}$ is non-increasing, bounded below and hence convergent. By an application of Taylor's theorem, we now obtain:

$$(4.2) \qquad \phi(x^k) - \phi(x^{k+1}) = \frac{1}{2} F_{i_k i_k}(y^k)(x_{i_k}^k - x_{i_k}^{k+1})^2, \quad k = 0,1,2,\ldots,$$

for some $y^k \in (x^k, x^{k+1})$ (where (x^k, x^{k+1}) denotes the open line segment joining x^k and x^{k+1}). Let $m = \text{glb}\{F_{i_k i_k}(y^k),\ k = 0,1,2,\ldots\}$. Since the Hessian of ϕ is positive semidefinite, we have $m \geq 0$. If $m = 0$, then by v) there exists some subsequence of $\{y^k\}$ converging to x^*. Since $\phi(y^k) \geq \phi(x^{k+1})$, $k = 0,1,2,\ldots$, it follows that $\phi(x^k) \to \phi(x^*)$ and thus $x^k \to x^*$.

We now consider the case when $m > 0$. Since $\{\phi(x^k)\}$ is convergent, we have from (4.2) that $\{x^k\}$ is Cauchy and thus converges to some $x \in S_\gamma$. We define the sets H_i, $i = 1,2,\ldots,n$, by $H_i = \{u \in S_\gamma : F_i(u) = 0\}$. It can be seen that the sets H_i are closed and nonempty. Further by the definition of $\{x^k\}$ it follows that for each $i = 1,2,\dots,n$ there exists a subsequence of $\{x^k\}$ in H_i which converges to x. This implies that $\tilde{x}$ is in H_i for all $i = 1,\dots,n$. Because of the uniqueness of the global minimum x^* it follows that $\tilde{x} = x^*$ and this completes the proof.

In the spirit of the above proof, we now consider the convergence of nonlinear SOR. Thus we consider, under the hypotheses of the above theorem, the iterative process which is generated by implementing one step of Newton's method instead of exact minimization in some coordinate direction as in the previous theorem. For the details of the proof we refer to [1].

<u>Theorem 4.2.</u> From any x^0 in S_γ, let $\{x^k\}$ be the sequence generated by

$$x_j^{k+1} = x_j^k, \quad j \neq i_k,$$

(4.3)

$$x_{i_k}^{k+1} = x_{i_k}^k - w_k \frac{F_{i_k}(x^k)}{F_{i_k i_k}(x^k)}.$$

Further let $\{I_k\}$ be defined by

$$I_k = \{x: \phi(x) \leq \phi(x^k) \text{ and } x_j = x_j^k, j \neq i_k\}$$

and let δ_k be defined by

$$\delta_k = \frac{F_{i_k i_k}(x^k)}{\max_{y \in I_k} F_{i_k i_k}(y)}.$$

Then the sequence $\{x^k\}$ is well-defined and converges to x^* if <u>either</u>

vi) there exists $\delta > 0$ such that $0 < \delta \leq w_k \leq 2\delta_k - \delta$, $k = 0,1,2,\dots$

<u>or</u> vii) there exists $g: R^n \to R$ such that g is continuous, nonnegative,

$g(v) = 0$ if and only if $v = 0$ and

$$0 \leq g(F(x^k)) \leq w_k \leq 2\delta_k - g(F(x^k)) \leq 2,\ k = 0,1,2,\ldots,$$

provided every coordinate direction has been chosen an infinite number of times in the above iteration process.

The proof of the above theorem is discussed via the decrease in the functional ϕ at each stage of the process. However as pointed out in Section 2, in the linear theory, one writes the SOR process as an iterative process and shows that the norm of this matrix L_ω is less than 1. One could proceed on similar lines also in the nonlinear case. Thus by repeated application of Taylor's theorem we can see [2] that for the nonlinear SOR problem

$$(4.4) \qquad x^{n+1} - x^* = |I - \omega_n (D_n - \omega_n L_n)^{-1} F'(\cdot)| (x^n - x^*)$$

where x^* is the unique minimum of ϕ. In the particular case when ϕ is a quadratic functional the above expression is exactly identical to the result stated in Section 2. D_n and L_n are diagonal and lower triangular matrices which change from step to step in the nonlinear SOR process.

We now state and outline another result on convergence of SOR for nonlinear problems which utilizes the linear theory extensively.

Theorem 4.3. For any arbitrary point $x_0 \in S_\delta, \delta \in (\phi(x^*), \gamma)$, there exists a sequence of relaxation parameters $\{\omega_n\}$, $0 < 0_n < 2$, such that the nonlinear SOR process (4.3) applied to the problem $F(x) = 0$ converges to x^*.

Outline of proof. For any starting point x_0 we can define the quadratic functional

$$\eta_0(x) = \frac{1}{2} < F'(x_0)(x-x_0), x-x_0 > + < F(x_0), x-x_0 > + \phi(x_0)$$

associated with the linear problem

$$(4.5) \qquad F'(x_0)(x-x_0) + F(x_0) = 0$$

whose solution will be denoted by $\bar{x}_0^*$.

Let $\bar{x}_1$ be the first step of the linear SOR applied to (4.5) and x_1 be the first step of the nonlinear SOR applied to $f(x) = 0$. Proceeding similarly we can generate $\{x_n\}$, $\{\bar{x}_n\}$, $\{\bar{x}_n^*\}$ and $\{\eta_n(x)\}$. We then show that

$$\phi(x_{n+1}) - \eta_n(\bar{x}_n^*) \leq k_n \; |\phi(x_n)-\eta_n(\bar{x}_n^*)|$$

where $k_n \in (0,1)$, by appropriate choice of ω. Then $\{\phi(x_n)\}$ can be shown to be nonincreasing and convergent. This leads us to note that $F(x_n) \to 0$ and $x_n \to x^*$. We have outlined here only the key steps of the proof and the details may be seen in [2].

In [8] the author discusses the problem of convergence of nonlinear SOR in a local sense. We now consider this question. Thus let $\omega, 0 < \omega < 2$, be fixed. The problem is to find a ball of radius r_0 around x^* such that the nonlinear SOR process with the same ω throughout the process converges to x^*. As noted earlier the nonlinear SOR process may be described by (4.4). From the linear theory, as in Section 2, we know that

$$\| I - (\omega D - \omega L)^{-1} A \|_{A^{\frac{1}{2}}} < 1$$

where $A = F'(x^*) = D - L - L^T$. Because of the appropriate continuity of ϕ, we can show that $D_n \to D$, $L_n \to L$ and thus we can show that for a suitably chosen ball around x^*, the operator

$$\| I - \omega(D_n - \omega L_n)^{-1} A_n \| \leq 1 - \varepsilon$$

for all n. This implies that the nonlinear SOR process converges for any arbitrary starting point x_0 chosen from this ball.

We conclude this section with a discussion on nonlinear SOR which is analogous to the result in Section 2 on linear SOR with varying relaxation parameters.

In the setting of Theorem 4.1, it can be seen that

$$\phi(x^k) - \phi(x^{k+1}) = \omega_k \left| \frac{F_{i_k i_k}(x^k)}{\overline{F}_{ik i_k}(\xi)} - \omega_k \right| \frac{\left|F_{i_k}(x_k)\right|^2}{\left|F_{i_k i_k}(x^k)\right|^2 \left|F_{i_k i_k}(\xi)\right|}$$

$$\geq \omega_k (2\delta_k - \omega_k) \frac{\left|F_{i_k i_k}(x_k)\right|^2}{\left|F_{i_k i_k}(x_k)\right|^2 \left|F_{i_k i_k}(\xi)\right|}$$

$$\geq 0.$$

Thus

$$\phi(x^0) - \phi(x^*) \geq \Sigma \left[\phi(x^n) - \phi(x^{n+1})\right]$$

$$\geq \Sigma\, \omega_k (2\delta_k - \omega_k) \frac{\left|F_{i_k}(x_k)\right|^2}{\left|F_{i_k i_k}(x_k)\right|^2 \left|F_{i_k i_k}(\xi)\right|}.$$

We now see that if x^k does not converge to x^*, then for at least one coordinate direction, say j, $\lim F_j(x^k) = m > 0$. But this implies that

$$\phi(x^0) - \phi(x^*) \geq \Sigma\, \omega_i^n (2\delta - \omega_j^n) m^2 / K^3$$

where K is an upper bound of $F_{jj}(x)$ over $S_{\phi(x^0)}$. Thus the hypothesis that $\Sigma\, \omega_j^n (2\delta - \omega_j^n)$ diverges leads to a contradiction that x^k does not converge to x^*. And this is a generalization of the corresponding result, Theorem 2.3, in Section 2.

Bibliography

[1] BREWSTER M.E., KANNAN R. - Nonlinear successive overrelaxation for nonlinear problems I. Numer. Math. 44 (1984), 309-315.

[2] BREWSTER M.E., KANNAN R. - Nonlinear successive overrelaxation for nonlinear problems II, Computing (to appear).

[3] CESARI L. - Sulla risoluzione dei sistemi di equazioni lineari per approssimazioni successive. Accademia Naz. Lincei, XXV (1937) 1-7.

[4] CESARI L. - Sulla risoluzione dei sistemi di equazioni lineari per approssimazioni successive. Rass, Poste, telegrafi e telefoni, Anno 9 (1937).

[5] CONCUS P. - Numerical solution of the minimal surface equation". Math. Comp. 21 (1967), 340-350.

[6] GREENSPAN D. - On approximating extremals of functionals". ICC Bulletin 4 (1965), 99-120.

[7] KELLER H.B. - Numerical solution of bifurcation and nonlinear eigenvalue problems. Application of Bifurcation Theory, P.H. Rabinowitz ed., Academic Press, New York 1977.

[8] LIEBERSTEIN H.M. - Overrelaxation for nonlinear elliptic partial differential equations, MRC Tech. Report *80, 1959.

[9] OSTROWSKI A.M. - On the linear iteration procedure for symmetric matrices, Rend. Math. Appl. 14 (1954) 140-163.

[10] RALL L.B. - Convergence of the Newton process to multiple solutions Numer. Math. 9 (1966), 23-37.

[11] SCHECHTER S. - "Iteration methods for nonlinear problems, Trans. Amer. Math. Soc. 104 (1962), 179-189.

[12] VARGA R.S. - Matrix Iterative Analysis, Prentice-Hall, 1962.

[13] WARGA J. - Minimizing certain convex functions. J. Soc. Indus. Appl. Math. 11 (1963), 588-593.

[14] YOUNG D.M. - Iterative solution of large linear systems, Academic Press, New York 1971.

WAVES IN WEAKLY-COUPLED PARABOLIC SYSTEMS

K. Kirchgässner
Universität Stuttgart
Mathematisches Institut A
Pfaffenwaldring 57 D-7000 Stuttgart 80

1. Introduction

The appearance of time periodic waves for weakly coupled parabolic systems received considerable attention in recent years. In particular, the widely spread interest in reaction-diffusion equations added to the popularity of this phenomenon. Numerous contributions have appeared (c. f. [3] for a survey). Nevertheless, a simple but general method seems to have escaped the attention of the authors. This method, which we shall explain here, yields the complete picture of bounded solutions near a trivial equilibrium. For reason of exposition we shall treat, pars pro toto, the following system proposed by Cohen, Hoppensteadt and Miura in [2]

$$\begin{aligned} u_t &= u_{xx} - \alpha u - \beta v - f_1(u,v) \\ v_t &= v_{xx} + \beta u - \gamma v - f_2(u,v) \end{aligned} \tag{1.1}$$

where an index denotes partial differentiation, α,β and γ are constants and the functions f_j are supposed to be real analytic in a neighborhood of 0 having a second order zero for $u = v = 0$. In [2] the authors considered special data for the constants and the functions f_j. They determined solutions $u(x - ct,t)$, $v(x-ct,t)$ for constant positive c and functions u,v which are bounded in the first and periodic in the second variable. They used a multiple-scaling method reminiscent of the WKB-method.

In this contribution we determine the complete solution-set near the zero equilibrium for a broader class of problems, indeed we describe a reduction process which works generally. It is only our lack of knowledge of some simple ODE-systems which cause the blank spots in our final survey of results. The method is based on a simple observation, namely that it is more appropriate to write (1.1) as an evolution equation in the x- than in the t-variable. Since we look for t-periodic solutions with given period, the corresponding differential operator has discrete spectrum, however extending to infinity in positive as well as in negative direction. Problems of this kind have been investigated recently in [4] and [6].

In order to emphasize the connection to the ODE-case we display the method in a way which has the spirit but not the details in common with the final approach. Define

$$\underline{u} = \begin{pmatrix} u_0 \\ u_1 \\ v_0 \\ v_1 \end{pmatrix}, \qquad \underline{f} = \begin{pmatrix} 0 \\ f_1 \\ 0 \\ f_2 \end{pmatrix},$$

$$A = \begin{pmatrix} 0 & 1 & 0 & 0 \\ \alpha+\partial_t & c & \beta & 0 \\ 0 & 0 & 0 & 1 \\ -\beta & 0 & \partial_t+\gamma & 0 \end{pmatrix}$$

then (1.1) can be rewritten as the following dynamical system

$$\frac{d\underline{u}}{dx} = A\underline{u} + \underline{f}(\underline{u}). \tag{1.2}$$

For periodic $\underline{u}$ with fixed period 2π (no loss of generality), the spectrum ΣA consists of isolated eigenvalues fo finite multiplicities. If ΣA does not intersect the imaginary axis $i\mathbb{R}$, the trivial solution is isolated in the space of bounded solutions.

The problem contains three natural parameters, namely the eigenvalues σ_1 and σ_2 of the matrix

$$A = \begin{pmatrix} \alpha & \beta \\ -\beta & \gamma \end{pmatrix}$$

and the wave velocity c. The eigenvalues ρ of $\mathcal{A}$ satisfy

$$(1.3) \qquad \rho^2 - c\rho = \sigma_j + ik \quad , \quad k \in \mathbb{Z} \, , \quad j = 1,2.$$

For large $|k|$ they behave like $\pm\{i|k|\}^{1/2}$. Therefore, ΣA lies in sectorial domains $|\arg\rho| < \omega < \pi/2$, $|\arg \rho - \pi| < \omega$ but is infinite in both. The idea is to separate a "critical" but finite part of Σ from the rest and to prove a center-manifold theorem. The critical part consists of those eigenvalues whose real parts are smallest in modulus. The separation-gap to the rest of the spectrum then determines the regularity of the center-manifold (CM).

In contrast to the known infinite-dimensional generalizations of the classical CM-theorem we cannot use the fact that the spectrum is finite **in the right** half of the complex plane (c.f. [5]). In effect a straight forward definition of semigroups for the positive and negative part of A via the usual Laplace integral fails and much deeper results are necessary (for elliptic systems see [4]). Here, however, we can use an ad hoc approach for the system (1.1).

The CM ~~is~~ finite dimensional and enherits all symmetry properties from the original equations. Moreover, it contains all small bounded solutions. Therefore, (1.2) can be reduced to a system of ordinary differential equations whose order coincides with the dimension of the CM, and thus with that of the critical eigenspace.

A case of particular mathematical interest arises when c=0 holds. Then (1.2) is reversible in the sense of G.D. Birkhoff, i.e. if R is the following matrix

$$R = \begin{pmatrix} a & & & 0 \\ & -a & & \\ & & a & \\ 0 & & & -a \end{pmatrix} , \quad a^2 = 1$$

then A and $\underline{f}$ satisfy

$$AR + RA = 0, \qquad \underline{f}(R\underline{u}) + R\underline{f}(\underline{u}) = 0.$$

This implies for a solution $\underline{u}(x)$, that $R\underline{u}(-x)$ is a solution as well. Moreover, ΣA is easily seen to be symmetric to the real and the imaginary axis. Thus, the critical eigenvalues appear in quadruples and higher degeneracies are to be expected than for $c \neq 0$.

Our method works for more than two equations in (1.1) also. Furthermore, the functions f_j may also depend on u_x and v_x as long as the growth is polynomial in these variables. Reversibility holds, if

$$f_j(au,-au_x,av,-av_x) = a\, f_j(u,u_x,v,v_x)$$

is satisfied for $j = 1,2$.

In section 2 we shall prove the CM-theorem, in section 3 the general situation is treated, first for $c = 0$, then for $c \neq 0$. In the first case we use for convenience instead of σ_1, σ_2 the two external parameters

$$\lambda = \alpha\gamma + \beta^2 \qquad , \qquad \mu = \alpha + \gamma$$

and obtain $\sigma^2 - \mu\sigma + \lambda = 0$. The qualitative behavior of solutions is different in the regions G_j shown in Figure 1

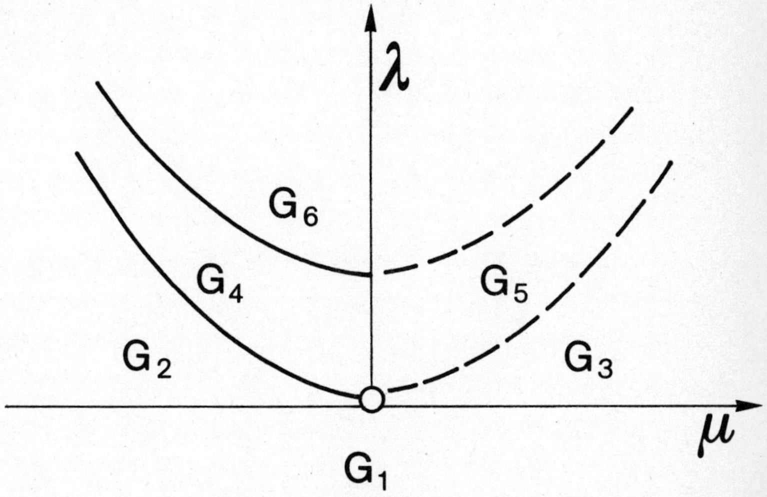

Figure 1

Along the dashed lines no bifurcation occurs. Therefore, the interesting transitions are $G_1 \to G_2$, $G_1 \to G_3$, $G_2 \to G_4$ and $G_4 \to G_6$. We restrict ourselves to non-generic cases, i.e. crossing through particular points are not considered. The only exception are those examples which were treated in [2]. The critical eigenvalues are given in the following Table 1

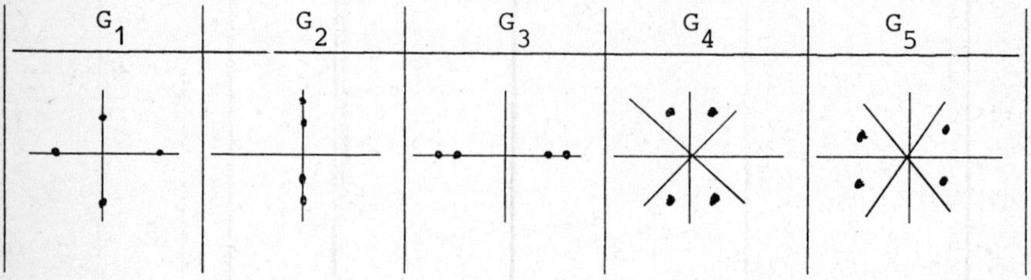

Table 1

The situation in G_6 corresponds to the one in G_4. While crossing the separating curve between G_4 and G_6 two pairs of simple eigenvalues penetrate the imaginary axis. The bifurcation phenomena can be resolved in some of the cases completely, in some, however, open problems remain.

For $c \neq 0$ the problem is much simpler. The bifurcation curves are obtained as follows by setting $\rho_r = 0$

$$
(1.3) \qquad \begin{aligned}
&c^4\lambda + c^2k^2\mu + k^4 = 0 && \text{for} \quad 4\lambda < \mu^2, \\
&2(k\pm\delta)^2 + c^2\mu = 0 && \text{for } \delta^2 = \lambda - \frac{\mu^2}{4} > 0 .
\end{aligned}
$$

If we change (λ,μ) along curves intersecting the curves described by (1.3) non-tangentially, then the critical eigenvalues transversally cross the imaginary axis. Therefore, if ρ does not go through zero, Hopf bifurcation occurs. Interesting phenomena happen when ρ crosses at 0. The detailed results are contained in section 3.

2. Reduction and Center-Manifold

The principle to reduce (1.1) resp. (1.2) to a finite system of ordinary differential equations is based on a generalized center-manifold (CM). This manifold is constructed for a finite part Σ_o of ΣA which is close to the imaginary axis. It is invariant under flows of (1.2) and contains all small bounded solutions; its regularity depends on the distance betwenn Σ_o and the rest of ΣA. We prove everything in this section in complex spaces and use for this purpose the natural complexification of A and $\underline{f}$ in (1.1). The results of interest which concern real solutions are obtained later by restriction.

The known generalization of CM-theorems do not work here as already indicated above. Therefore, we apply an ad-hoc approach which has been used for a scalar equation already in [6]. Equation (1.1) is treated as an evolution equation in

$$H^k_{\#} = \{u \in (H^k_{loc}(\mathbb{R}))^2 / u(x+2\pi) = u(x) \text{ a.e.}\}.$$

We henceforth suppress the underlining of vector-variables. $H^k(\Omega)$ denotes the usual complex Sobolev-space $W^k_2(\Omega)$. The scalar product in $H^o_{\#}$ is given by

$$(u,v) = \frac{1}{2\pi} \int_o^{2\pi} u(t) \cdot v(t) dt ,$$

$$u \cdot v = \sum_{j=1}^{2} u_j \bar{v}_j .$$

For the norm in $H^k_{\#}$ we obtain

$$|u|_k = \{\sum_{j=o}^{k} (\partial_t^j u, \partial_t^j u)\}^{1/2}.$$

Observe that $u \in H^k_{\#}$ if and only if

$$\sum_{\nu\in\mathbb{Z}} |\nu|^{2k} |u^\nu|^2 < \infty \quad , \quad u^\nu = (u, e^{i\nu t}) .$$

This equivalence extends to non-integer k.

Let $C^k(\mathbb{R}\, H)$ denote the space of k-times continuously differentiable functions from $\mathbb{R}$ to the normed space $(H, \|\cdot\|)$ endowed with the usual Fréchet-structure generated by the seminorms

$$p_K^j(f) = \sup_{x\in K} \| \partial_x^j f(x) \| \quad , \; j = 0,\dots,k,$$

$K\subset\mathbb{R}$ compact. $C_b^k(\mathbb{R}, H)$ is the normed subspace of bounded functions with bounded derivatives up to the order k. The complete metric subspace of C_b^k whose functions satisfy

$$\sup_{x\in\mathbb{R}} \| \partial_x^j f(x) \| \leq \delta \; , \; j = 0,\dots,k$$

is denoted by $C_\delta^k(\mathbb{R}, H)$. We need the following particular spaces

$$Y^o = C^o(\mathbb{R}, H_\#^o) ,$$

$$X^l = C^l(\mathbb{R}, H_\#^{1-l/2}) \quad , \quad l = 0,1,2 ,$$

$$Z^l = X^o \cap X^l .$$

The norms in Y_b^o resp. Z_b^l are denoted by $\|\cdot\|^o$ resp. $\|\cdot\|_l$ i.e.

$$\|u\|^o = \sup_{x\in\mathbb{R}} |u(x)|_o ,$$

$$\|u\|_l = \sup_{x\in\mathbb{R}} \{|u(x)|_1 + |u''(x)|_o\}$$

where $'' = \partial_x^2$.

For $u \in Z^2$ we consider (1.1) in the form

(2.1) $$Lu = \partial_x^2 u - c\,\partial_x u - (\partial_t + A)u = f(u)\,.$$

<u>Assumption 2.1</u>: We suppose that, for some $m \geq 2$, $\delta_1 > 0$,

$$f \in C_b^m(H_\#^1, H_\#^0) \cap C_{\delta_1}^1(X_\#^1, H_\#^0)$$

holds.

Observe that the regularity of f is trivial for every m as long as f_1 and f_2 are real analytic in u and v. The boundedness and smallness requirements can be achieved by multiplying with a cutoff-function which is 1 in a neighborhood of 0 (depending on δ_1) and 0 outside a slightly larger neighborhood. Since the original f coincides with the f considered here only near 0, our results are local in nature and are restricted to solutions small in $H_\#^1$.

Next we split ΣA as follows

$$\Sigma = \Sigma_0 \cup \Sigma_1 \quad , \quad \Sigma_0 \cap \Sigma_1 = \emptyset$$

where the critical part Σ_0 is finite. For the regularity of CM this splitting has to be sufficiently strong. Set $\rho = \rho_r + i\rho_i$,

$$a_0 > \sup\{|\rho_r| \;/\; \rho \in \Sigma_0\}\,,$$
$$a_1 < \inf\{|\rho_r| \;/\; \rho \in \Sigma_1\}\,,$$

then we suppose that

(2.2) $$m a_0 < a_1$$

holds. Furthermore, we assume that $\sigma_1 \neq \sigma_2$. According to (1.3) the eigenvalues of A occur for c=0 at the corners of a square with center at 0. Therefore, Σ_0 is determined by a finite subset ($n = |S_0|$)

$$S_0 = \{(j,k)\} \subset \{1,2\} \times \mathbb{Z} = S\,.$$

For $c \neq 0$ the points in Σ_o may also depend on the root of (1.3) even for given j and k. But we suppress the necessary but minor modifications.

The partition of Σ generates a decomposition of all spaces considered, e.g.

$$H^k_{\#} = H^k_{\#,0} \oplus H^k_{\#,1} ,$$

$$X^l = X^l_o \oplus X^l_1 \qquad , \quad l = 0,1,2 ,$$

and similarly for Z^ℓ, Y^o and for the operator L. $H^k_{\#,0}$ has dimension $n = |S_o|$ and may be identified with $\mathbb{C}^n$. The decomposition is constructed explicitly via the eigenvectors φ_j of the matrix A corresponding to σ_j, $S_1 = S - S_o$

$$u_r = \sum_{(j,k)\in S_r} u_r^{k,j} \varphi_j e^{ikt} \qquad , \quad r = 0,1 .$$

The function u belogns to $H^\ell_{\#}$ if and only if

$$\sum_{(j,k)\in S_1} |k|^{2l} |u_1^{k,j}|^2 < \infty .$$

Subsequently we write u_r^k instead of $u_r^{k,j}$ and thus suppress the dependence on j.

<u>Lemma 2.1</u>: For $f \in Y^o_{1,b}$ there exists a unique solution $u_1 \in Z^2_{1,b}$ of $L_1 u_1 = f_1$. Moreover, the following estimate holds

$$\|u_1\|_l \leq \gamma_1 \|f\|^o \qquad , \quad l = 0,1,2 .$$

<u>Proof</u>: The equation $L_1 u_1 = f_1$ reads for the Fourier components $u_1^k = u_1^{k,j}$

$$u_1^{k''} - c u_1^{k'} - (ik + \sigma_j) u_1^k = f_1^k \tag{2.3}$$

for $(j,k) \in S_1$. Since $\rho_r^k \neq 0$ a Green's function of (2.3) exists for bounded solutions. Denote it by $G^k(x-\xi)$. Set a_1^k to be the minimum of $|\rho_r^{k,j}|$

then, since $\rho^k \sim (i|k|)^{1/2}$ for large $|k|$, the following estimates hold

$$|G^k(x)| \leq \frac{c}{\sqrt{|k|}} e^{-a_1^k|x|},$$

$$|G^{k'}(x)| \leq c\, e^{-a_1^k|x|}.$$

Therefore, the solution of (2.3) is given by

$$u_1^k(x) = \int_{-\infty}^{\infty} G^k(x-\xi) f_1^k(\xi)\, d\xi .$$

In view of $a_1^k = O(|k|^{1/2})$, $k\, u_1^k \in \ell_2$, $k^{1/2} u_1^k \in \ell_2$ and (2.3) yields $u_1 \in X_b^\ell$ for $\ell = 0,1,2$. Moreover, we have for $\ell = 0$ and 1

$$\|u_1\|_\ell^2 = \sup_{x \in IR} \sum_{k \in Z} |u_1^k(x)|^2 |k|^{2-\ell} \leq c^2 (\|f_1\|^o)^2,$$

Using (2.3) one obtains the inequality as required.

<u>Remarks 2.2</u>

1. The unique solution of $L_1 u_1 = f_1$ has the representation

(2.4) $$u_1(x) = \int_{-\infty}^{\infty} K_1(x-\xi) f_1(\xi) d\xi$$

where K_1 is given by its Fourier-components $G^k = G^{k,j}$.
Moreover, K_1 as a mapping from $Y_{1,b}^o$ to $Z_{1,b}^2$ is bounded.

2. The assertion of Lemma 2.1 remains true if only $f_1 e^{a|x|} \in Y_{1,b}^o$ for some $0 < a < a_1$.

The CM is sought in the form

$$u_1 = h(u_o, u_o')$$

where the arguments of h can be identified with $\mathbb{C}^{2n} = \mathbb{C}^n \times \mathbb{C}^n$. For this reason we define $\mathcal{H}$ as follows

$$(2.5) \qquad H = \mathbb{C}^1_\delta\,(\mathbb{C}^{2n}, H^1_\#) \cap C^m_b\,(\mathbb{C}^{2n}, H^1_\#)$$

for fixed $m \in N$ (see assumption (2.1.)) and $\delta > 0$ of which we dispose in the course of the proof.

Lemma 2.3: Assume $h \in H$, let f satisfy assumption (2.1). Consider the initial value problem

$$(2.6) \qquad \begin{aligned} &L_o u_o = f_o(u_o + h(u_o, u_o')), \\ &u_o(0) = \xi_o \quad , \quad u_o'(0) = \xi_1 \quad , \quad \xi = (\xi_o, \xi_1) . \end{aligned}$$

Then (2.6) has a unique solution $u_o \in Z^2$ with $u_o \in C^m(\mathbb{C}^{2n} \times H, Z_o^2)$. Furthermore, the following estimates are valid

$$|u_o(x)|_1 + |u_o'(x)|_1 \leq \gamma_2 \; (|\xi| + ||f_o||_x) e^{a_o|x|}$$

$$|D^\alpha_{(\xi,h)} u_o(x)| + |D^\alpha_{(\xi,h)} u_o'(x)| \leq \gamma_2 \; e^{|\alpha| a_o |x|}$$

where $\alpha \in N_o^2$, $|\alpha| = \alpha_1 + \alpha_2 \leq m$. Here we have used the following abbreviation

$$||f_o||_x = \sup_{|y| \leq x} |f(y)|_o .$$

Proof: Since (2.6) is a finite-dimensional system the proof is standard (c.f. [1]). Nevertheless, we indicate the necessary steps for completeness. Define the nxn-matrices U_j by the following initial value problem $(j = 1,2)$

$$\begin{aligned} L_o U_j = 0 \quad , \quad &U_o(0) = I \quad , \; U_o'(0) = 0 , \\ &U_1(0) = 0 \quad , \; U_1'(0) = I , \end{aligned}$$

I is the unit matrix in $\mathbb{C}^n$. The solution of (2.6) can be written as

$$u_o(x) = U_o(x)\xi_o + U_1(x)\xi_1 + \int_o^x U_1(x-t)f_o(t)dt \ .$$

Obviously, $e^{-a_o|x|}U_j(x)$ is bounded in $\mathbb{R}$. The uniqueness of u_o is trivial. The smoothness with respect to h and ξ follows by a simple application of the implicit function theorem. The above representation immediately yields the first inequality. The remaining estimates follow by successive differentiation and using assumption 2.1 for sufficiently small δ_1. The $|\alpha|$th. derivative involves the $|\alpha|$th. power of first order derivatives as the fastest growing term in $|x|$. Therefore, the exponential growth increases with $|\alpha|a_o$.

Lemma 2.4: Assume (2.1) for f. There exists a $\delta_1 > 0$ such that for all $\xi \in \mathbb{C}^{2n}$

$$Lu = f(u) \quad , \quad u_o(0) = \xi_o \quad , \quad u_o'(0) = \xi_1$$

has at most one solution in Z_b^2.

Proof: Let $u, \tilde{u} \in Z_b^2$ be two solutions, define $d = u - \tilde{u}$, then

$$|f(\tilde{u}(x)) - f(u(x))|_o \leq \delta_1 |d(x)|_1 \ .$$

Apply Lemma (2.3) and obtain

$$(2.7) \qquad |d_o(x)|_1 \leq \gamma_2 \delta_1 e^{a_o|x|} \quad (\|d_o\|_o + \|d_1\|_o) \ .$$

Since $d \in Z_b^2$, we have $|d(x)|_1 \leq \|d\|_o$. Hence Lemma (2.1) implies

$$(2.8) \qquad \|d_1\|_o \leq \delta_2 \|d_o\|_o \quad , \quad \delta_2 = \frac{\gamma_1\delta_1}{1-\gamma_1\delta_1} \ .$$

Taking the supremum on the left of (2.7) we obtain

$$\|d_o\|_o \leq \gamma_2\delta_1(1+\delta_2)e^{\ a_o|x|}\|d_o\|_o \ .$$

Therefore, $d_o(x) = 0$ for some interval $[-h,h]$ containing 0. Hence by (2.8), $d_1(x) = 0$ in the same interval, and thus $u = \tilde{u}$ holds there. Now define $u_h(x) = u(x+h)$, $\tilde{u}_h(x) = \tilde{u}(x+h)$, then the same argument yields $u_h = \tilde{u}_h$ in $[-h,h]$. Thus we have extended the uniqueness to $[-h,3h]$. Continuing the same argument indefinitely yields the assertion.

Theorem 2.5

Suppose that assumption (2.1) holds for f. Moreover let (2.2) be satisfied. Then, for sufficiently small positive δ and δ_1, there exists a unique $h \in H$ such that the manifold

$$M = \{u \in Z^2 \;/\; u_1(x) = h(u_o(x), u_o'(x))\}$$

is invariant under (2.1) if u_1 is bounded. Moreover, $u \in Z_b^2$ implies $u \in M$.

Proof: It is readily seen from (2.4) that h, if it exists, must satisfy

$$h(\xi) = Th\,(\xi)$$

where

(2.9) $$(Th)(\xi) = \int_{-\infty}^{\infty} K_1(-t) f_1(u_o(t;\xi,h) + h(u_o,u_o')(t;\xi,h))dt.$$

The mapping T acts in $C_\delta^1(C^{2n}, H_{\#}^1)$. By Lemma 2.2 and Lemma 2.3, Th is a C^m-function of ξ. Moreover, we have

$$|(Th)(\xi)|_1 \le \gamma_1 \delta_1 \le \delta .$$

The estimates of Lemma 2.3 together with (2.2) show that differentiation can be interchanged with the integral. Then we obtain

$$|D_\xi Th(\xi)|_1 \le \delta_1 \int_{-\infty}^{\infty} e^{-a_1|t|}\{|D_\xi u_o(t)|_o(1+\delta)$$

$$+ \delta |D_\xi u_o'(t)|_o\}dt$$

$$\leq c\delta_1 \int_{-\infty}^{\infty} e^{(a_0-a_1)|t|} dt \leq c'\delta_1 \leq \delta .$$

Furthermore, T is continuously differentiable with respect to h. This can be shown as above by differentiating under the integral. For sufficiently small δ_1 we obtain

$$|D_h T(h)(\xi)| \leq \delta .$$

Therefore, T is a contraction provided $\delta < 1$ holds. The unique fixed point $h \in C^1_\delta(\mathbb{C}^{2n}, H^1_{\#})$ is the required solution. Its regularity follows by successive differentiation of (2.9) up to the order m.

The invariance of M: assume u solves (2.1), u_1 bounded,

$$u_1(0) = h(\xi) \quad , \quad u_1'(0) = \partial_{u_0} h(\xi) u_0'(0) + \partial_{u_0'} h(\xi) u_0''(0) .$$

It follows that $u_0(x)+h(u_0(x),u_0'(x)) = v(x)$ solves (2.1) in Z^2, since

$$h(u_0(x),u_0'(x)) = \int_{-\infty}^{\infty} K_1(x-t) f_1(u_0(t;\xi)$$
$$+ h(u_0,u_0')(t;\xi))dt$$

holds in view of (2.4). Moreover, $h(u_0,u_0')(x)$ is bounded in $H^1_{\#}$, v and u must coincide by Lemma 2.4.

Finally, if $u \in Z^2_b$ solves (2.1) set $\xi=(u_0(0),u_0'(0))$ and $u_0(x)+h(u_0,u_0')(x)= =v(x)$. By Lemma 2.4 we have u=v, qed.

Corollary 2.6

1. Assume $D^j_u f(0)=0$ for $j=0,\dots,\ell\leq m-1$, then $D^j_\xi h(0) = 0$.
2. For c=0 we have

$$h(\xi_0,-\xi_1) = h(\xi_0,\xi_1) . \tag{2.10}$$

Proof: Restrict H to

$$\overset{o}{H}_{\ell} = \{h \in H \;/\; D^{j}h(0) = 0 \quad , \quad j = 0,\ldots, \ell-1 \;,$$

If $h \in \overset{o}{H}_{\ell}$, $D^{j}_{\xi}u_{o}(\cdot;0,h) = 0$ by the unique solvability of (2.6). Moreover it follows from (2.9) that $D^{j}_{\xi}(Th)(0) = 0$ holds. Thus T acts in $\overset{o}{H}_{\ell}$ and the existence of a unique fixed point in $\overset{o}{H}_{\ell}$ can be derived as in the proof of Theorem 2.5.

For the proof of 2. assume that u_{o} solves (2.6). Then $v_{o}(x)=u_{o}(-x)$ satisfies

$$L_{o}v_{o} = f_{o}(v_{o} + h(v_{o},v_{o}')) \;,$$

$$v_{o}(0) = \xi_{o} \quad , \quad v_{o}'(0) = -\xi_{1} .$$

Observe further that in this reversible case $K_{1}(-x) = K_{1}(x)$.

Therefore, restricting H to

$$H_{s} = \{h \in H \;/\; h \text{ satisfies } (2.10)\}$$

implies that v_{o} solves (2.6) for $\xi = (\xi_{o},-\xi_{1})$ and that $Th(\xi_{o},-\xi_{1}) = Th(\xi_{o},\xi_{1})$ from (2.9). Thus T maps H_{s} into itself and the assertion follows.

3. Reduced Equations and Bifurcation

In this section the bounded solutions $(u \in Z^{2}_{b})$ of (2.1) are determined in the regions G_{j} of figure 1. Using the results of the preceding section we reduce (2.1) to a finite system of ordinary differential equations which are finally discussed as far as possible.

First we take $f(u)=(f_{1}(u,v), f_{2}(u,v))$ from section 1, real analytic and having a second order zero at u=v=0. Since $C^{o}_{\#}$ is imbedded in $H^{1}_{\#}$, f complexified is an analytic mapping from $H^{1}_{\#}$ to $H^{o}_{\#}$. Multiply f by some cut-off - function $\chi \in C^{\infty}(\mathbb{R})$

$$\chi(|u|) = \begin{cases} 1 \quad , |u| \le \varepsilon \\ 0 \quad , |u| \le 2\varepsilon \ , \end{cases}$$

$$\varepsilon \, \chi'(|u|) \le C$$

then

$$|(\chi f)(u)|_o + |(\chi f)'(u)|_o \le C'\varepsilon \le \delta$$

for all $u \varepsilon H^1_{\#}$ if $\varepsilon > 0$ is sufficiently small. Thus χf satisfies assumption (2.1) and replacing f in (1.1) by χf restrict our results to some neighborhood of $u=v=0$. Henceforth we do not explicity mention this local character of our results.
Moreover, we write f in place of χf.
The CM of Theorem 2.5 reduces the search for Z^2_b-solutions of (2.1) to

$$L_o u_o = f_o(u_o + h(u_o, u_o')) \tag{3.1}$$

Where $h \in \overset{o}{H}_l$ for at least $l=2$. The vector-function u_o lives on the eigenspace corresponding to the "critical" eigenvalues ρ of A, i.e. $\rho \in \Sigma_o$. We restrict our consideration to generic cases, when the G_j-dividing curves are crossed transversally and special points are excluded. The cases treated in [2] however are nongeneric. We therefore break our rule and discuss these results as well.

<u>Lemma 3.1</u>: Let $|\rho_r| \ge a > 0$ hold for some a and all $\rho \in \Sigma A$. Then there exists an a-dependent neighborhood U of 0 in Z^2_b such that 0 is the only solution of (2.1) in U.

Proof: Use Lemma 2.1. Since $f=f_1$ we obtain for every solution $u \in Z^2_b$

$$\|u\|_2 \le \gamma_1 \|f\|^o \le \gamma_1 \delta_1 \|u\|_2 \, .$$

Therefore, if $\gamma_1 \delta_1 < 1$, $u = 0$ follows. But γ_1 depends on a and thus

$\delta_1 \to 0$ as $a \to 0$; qed.

This Lemma implies that bifurcation from 0 can occur only if ΣA has points in common with $i\mathbb{R}$. In this connection the reversible case is of particular interest. Therefore, we treat the case $c=0$ first.

$G_1(\lambda<0, \mu \in \mathbb{R})$: Here $\sigma \in \mathbb{R}$ and the critical eigenvalues must satisfy $\sigma^2=\sigma_j$, $j=1,2$, i.e. $k=0$. Moreover $\sigma_1\sigma_2 < 0$ and therefore $\pm i(-\sigma_1)^{1/2}$ are critical. We have $n=2$ and the reduced equation reads

$$u_o'' + \sigma_1 u_o - f_o(u_o, h(u_o, u_o')) = 0 \,. \tag{3.2}$$

The function u_o and f_o are scalar-valued and can be taken as real. Classical bifurcation theory applies. Set $U_o(\omega x)=u_o(x)$ and consider the space of even, 2π-periodic functions, then (3.2) reads

$$\omega^2 U_o'' + \sigma_1 U_o - f_o(U_o + h(U_o, \omega U_o')) = 0 \,. \tag{3.2)'$$

Its linear part has a 1-dimensional kernel for $\omega_o=(-\sigma_1)^{1/2}$ and its range has codimension one. Simple eigenvalue-bifurcation applies in the (ω, U_o)-space. Since every solution u_o generates a one-parameter family of solutions $u_c(x)=u_o(x+c)$ it can be seen that all bounded solutions are obtained.

Proposition 3.2: For each $(\lambda,\mu) \in G_1$, every real and bounded solution of (2.1) is periodic in x.

$G_2(0<\lambda<\frac{\mu^2}{4}, \mu<0)$: Both σ_j are negative. The critical eigenvalues are $\pm i(-\sigma_j)^{1/2}$ and thus $n=4$. The reduced equations read

$$\begin{aligned} u_o'' + \sigma_1 u_o - f_{o1}((u_o, v_o) + h(u_o, v_o, u_o', v_o')) &= 0 \,, \\ v_o'' + \sigma_2 v_o - f_{o2}((u_o, v_o) + h(u_o, v_o, u_o', v_o')) &= 0 \,. \end{aligned} \tag{3.3}$$

Again u_o, v_o, f_{o1}, f_{o2} are scalar, real-valued functions. Obviously, quasi periodic solutions are to be expected. We revoke results of J. Moser

[7] and also J. Scheurle [8] for the case where f_o and h are analytic. Set $\omega_j = (-\sigma_j)^{1/2}$ and consider λ and μ as parameters. We have to demand

$$\frac{\partial(\omega_1,\omega_2)}{\partial(\lambda_o,\mu_o)} = \frac{\sigma_1 - \sigma_2}{\sqrt{\lambda_o}(2\sigma_1-\mu_o)(2\sigma_2-\mu_o)} \tag{3.4}$$

which holds everywhere in G_2. Moreover, one needs the following non-resonance condition

$$|j_1\omega_1(\lambda_o,\mu_o) + j_2\omega_2(\lambda_o,\mu_o)| \geq \frac{\gamma}{|(j_1,j_2)|^{\tau}} \tag{3.5}$$

for all $(j_1,j_2) \in \mathbb{Z}^2-\{0\}$ and some positive constants γ and τ.
Given (3.4) and (3.5), (λ_o,μ_o), u=0 is a bifurcation point of quasiperiodic solutions with two independent frequencies.

For our purpose the analyticity requirement is too strong for f and h. One would need the above result for C^m-nonlinearities, m sufficiently large. Since such a result seems not yet to be available, we state this part as a conjecture. The question what happens in the exceptional points where (3.5) is violated remains also open.

A much weaker result can be obtained under the condition that ω_1 is not an integer multiple of ω_2 or vice versa. Then we can proceed as in G_1 and obtain periodic solutions of period $2\pi/\omega$, $\omega\sim\omega_1$ or $\omega\sim\omega_2$. However, since σ_j depends analytically on λ and μ, the above condition holds in an open and dense set.

<u>Proposition 3.3</u>: There is an open and dense set $\Omega\subset G_2$ such that for each $(\lambda,\mu) \in \Omega$ there exist nontrivial x-periodic real solutions of (2.1). Moreover we conjecture: Given the conditions (3.4) and (3.5), u=0, $\lambda=\lambda_o$, $\mu=\mu_o$ is a bifurcation point of real, quasiperiodic solutions with two independent frequencies.

The transition $G_1\to G_2$ is not yet resolved. The reduced equations coincide with **(3.3)**, but σ_2 changes sign. For special assumptions one can prove the existence of homoclinic or heteroclinic orbits in the same spirit as below for G_3. However, these conditions are not invariant under linear transformations of the phase space and therefore of no relevance.

$G_3\,(0<\lambda<\mu^2/4,\ \mu>0)$: In G_3 the trivial solution is isolated. However, the transition $G_1 \to G_3$ is of interest. Here, σ_1 changes sign and σ_3 is positive. The critical eigenvalues are $\pm i(-\sigma_1)^{1/2}$ in G_1 and $\pm(\sigma_1)^{1/2}$ in G_3. The reduced equation reads (n=2)

$$u''_o + \sigma_1 u_o - f_o(u_o + h(u_o,u'_o)) = 0 \tag{3.6}$$

u_o, f_o being scalar- and real-valued. Assume $|f(u)| = O(|u|^s)$ and

$$f_o(u_o) = a_s u_o^s + O(u_o^{s+1}),\ a_s \neq 0,\ s \geq 2\,. \tag{3.7}$$

According to Corollary 2.6 we have

$$f_o(u_o + h(u_o,u'_o)) = a_s u_o^s + O(u_o^{s+1})\,. \tag{3.8}$$

The truncated equation

$$u''_o + {}_1u_o - a_s u_o^s = 0$$

has a saddle node in G_3 and a center in G_1 at $u_o = u'_o = 0$. For even s one further equilibrium exists being a center in G_3 and a saddle in G_1. A homoclinic orbit bifurcates from 0 at the separation line between G_1 and G_3. For odd s the situation depends on the sign of a_s. The results are shown in Figure 2.

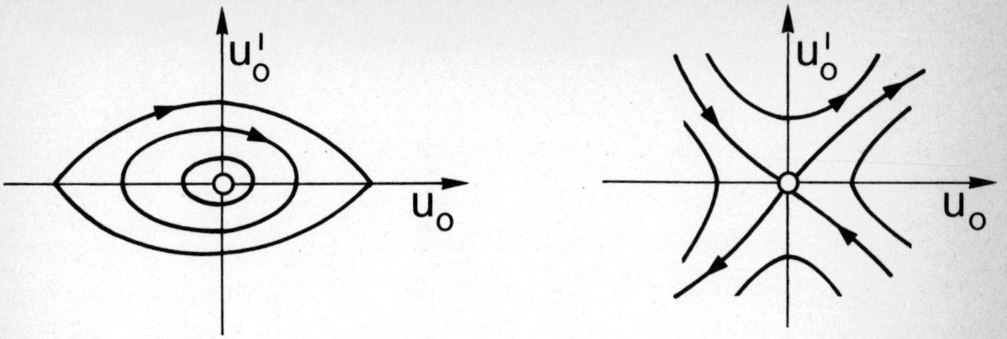

a) $a_s > 0$, $(\lambda,\mu) \in G_1$ b) $a_s > 0$, $(\lambda,\mu) \in G_3$

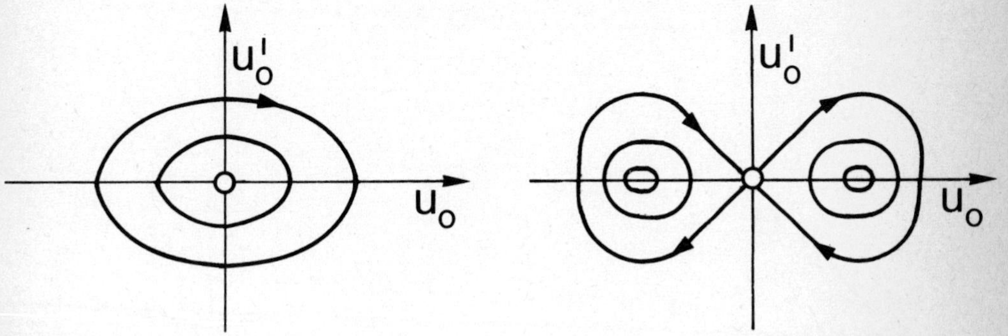

c) $a_s < 0$, $(\lambda,\mu) \in G_1$ d) $a_s < 0$, $(\lambda,\mu) \in G_3$

Figure 2

These pictures prevail locally for the full equations (3.8) (for s odd, $\underline{f}$ should be odd). The proof follows from the observation that centers are stable against reversible perturbations. Details can be found in [6]. The equilibria in Figure 2 are x-independent $H^1_{\#}$-solutions of (2.1) when considered on the CM. We speak of homoclinic resp. heteroclinic solutions u of (2.1) if $u \in Z_b^2$ and

$$\lim_{|x|\to\infty} u(x) = u_\infty \qquad \text{resp.} \qquad \lim_{x\to\pm\infty} u(x) = u_\pm$$

exist, and if u_∞ resp. $u_\pm$ solve (2.1) in $H^1_\#$.

Proposition 3.5: Assume that (3.7) holds. Take any point $\mu > 0$, $\lambda = 0$ on the separating line of G_1 and G_3. Then (i) if s is even, $a_s \neq 0$, $(\mu, 0, u=0)$ is a bifurcation point of periodic and of homoclinic real solutions of (2.1); (ii) if s and f are odd, $a_s \neq 0$, $(\mu, 0, u=0)$ is a bifurcation point of periodic and of heteroclinic real solutions of (2.1).

G_4 $(\mu < 0,\ 0 < \lambda - \frac{\mu^2}{4} < 1$ and beyond: The critical eigenvalues are situated on the hyperbola $\rho_r^2 - \rho_i^2 = \mu/2$. Since $2\rho_r\rho_i \neq 0$, 0 is an isolated solution in G_4. Of interest are the transitions $G_2 \to G_4$ and $G_4 \to G_6$. The eigenvalue behavior is shown in Figure 3.

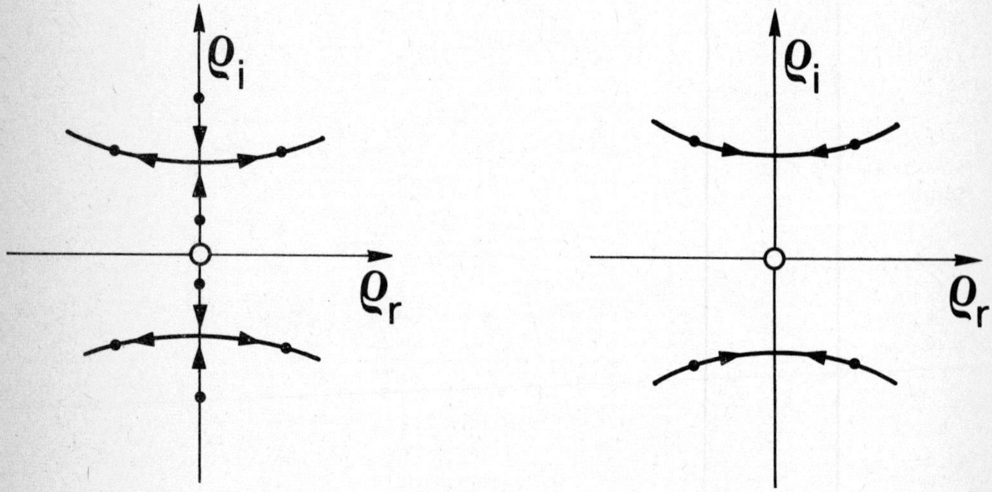

Figure 3

For $G_4 \to G_6$ the eigenvalues cross the imaginary axis while, for $G_2 \to G_4$, they meet on $i\,\mathbb{R}$ and then leave the imaginary axis. In both cases I can only handle special problems and am still far from systematic results. Similar situations as in b) occur if the parabolas $\lambda - \frac{\mu^2}{4} = k^2$ are crossed, $k \in \mathbb{N}$, $k \geq 1$.

The case $c \neq 0$

The eigenvalues ρ are determined as roots of $\rho^2 - c\rho - (\sigma + ik) = 0$. For small $|\rho_r|$ we have

$$\rho_r = -\frac{1}{c}(\rho_1^2 + \sigma_r) + \dots ,$$

$$\rho_i = -\frac{1}{c}(\sigma_i + k) \quad + \dots .$$

For the discussion we find it appropriate to use $\tau = \lambda - \mu^2/4$ instead of λ.

Consider $\tau < 0$: Then $\sigma_i = 0$ and $\rho_i = -k/c$. Therefore, $\rho_r = 0$ if and only if

$$(3.9) \qquad \left(\frac{k^2}{c^2} + \frac{\mu}{2}\right)^2 + \tau = 0 .$$

If we cross one of these parabolas non-tangentially, then $\rho_r' \neq 0$, i.e. ρ crosses the imaginary axis with non-vanishing velocity.

For $k = 0$ we have $\rho_i = 0$. Then $\rho_r = 0$ implies $\lambda = 0$. Therefore, we have to investigate $\lambda \approx 0$. It is easy to see that one critical eigenvalue exists

$$\rho = -\frac{\lambda}{c\mu} + O(\lambda^2) .$$

The reduced equation is real and of first order; it reads

$$(3.10) \qquad u_o' + \frac{\lambda}{c\mu} u_o - f_o(u_o + h(u_o)) = 0 .$$

Assume (3.7) for f_o. Then (3.10) has the equilibria

$$u_o = 0 \quad , \quad u_o = \left(\frac{\lambda}{c\mu a_s}\right)^{1/(s-1)} .$$

Both are stable with respect to higher order perturbations. Therefore, the solution set of (3.9) looks as shown in Figure 4 ($a_s > 0$, $\frac{\lambda}{\mu} > 0$).

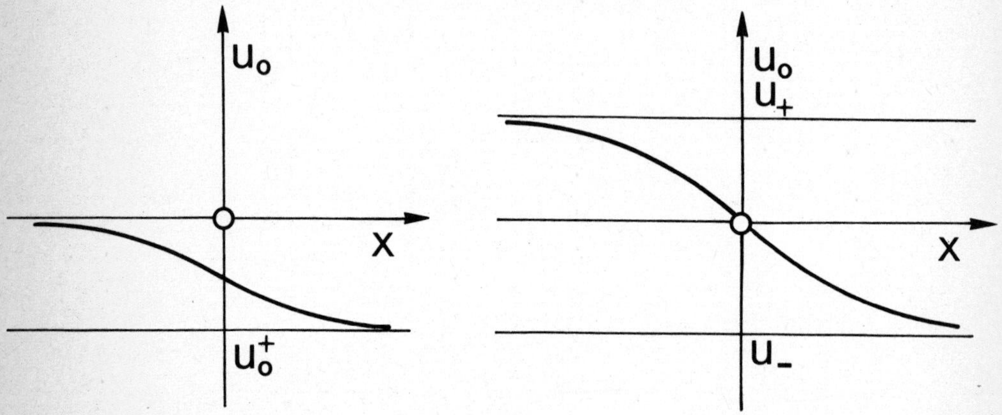

a) s even

b) s odd

Figure 4

For $k \neq 0$ we have $\rho_i = -k/c \neq 0$. If we require

$$(3.11) \qquad \tau + \frac{k^2 - l^2}{4c^2} = 0 \quad , \quad (k,l) \in N^2 \quad , \quad k \neq l$$

resonance cannot occur. Therefore, we have Hopf-bifurcation.

Proposition 3.6: Assume $\tau < 0$ and suppose that (3.7) holds.
(i) $k = 0$: Let (τ,μ) be any point on the parabola (3.9), then $(\tau,\mu,u=0)$ is a bifurcation point of heteroclinic orbits.
(ii) $k \neq 0$: Let (τ,μ) be a point on the corresponding parabola, then $(\tau,\mu,u=0)$ is a point of Hopf-bifurcation if (3.11) holds.

Now consider $\tau > 0$: We have $\sigma_i = \pm\sqrt{\tau}$ and $\sigma_r = \mu/2$. Therefore, $\rho_r=0$ if

(μ,τ) satisfy

$$(3.12) \qquad (\tau-k^2 + \frac{\mu c^2}{2})^2 + 2c^2k^2\mu = 0 .$$

These are, for $k \neq 0$, parts of parabolas with axis parallel to $\tau = -\mu$. They touch the τ-axis in k^2. Figure 5 shows the critical curves (3.9) and (3.12).

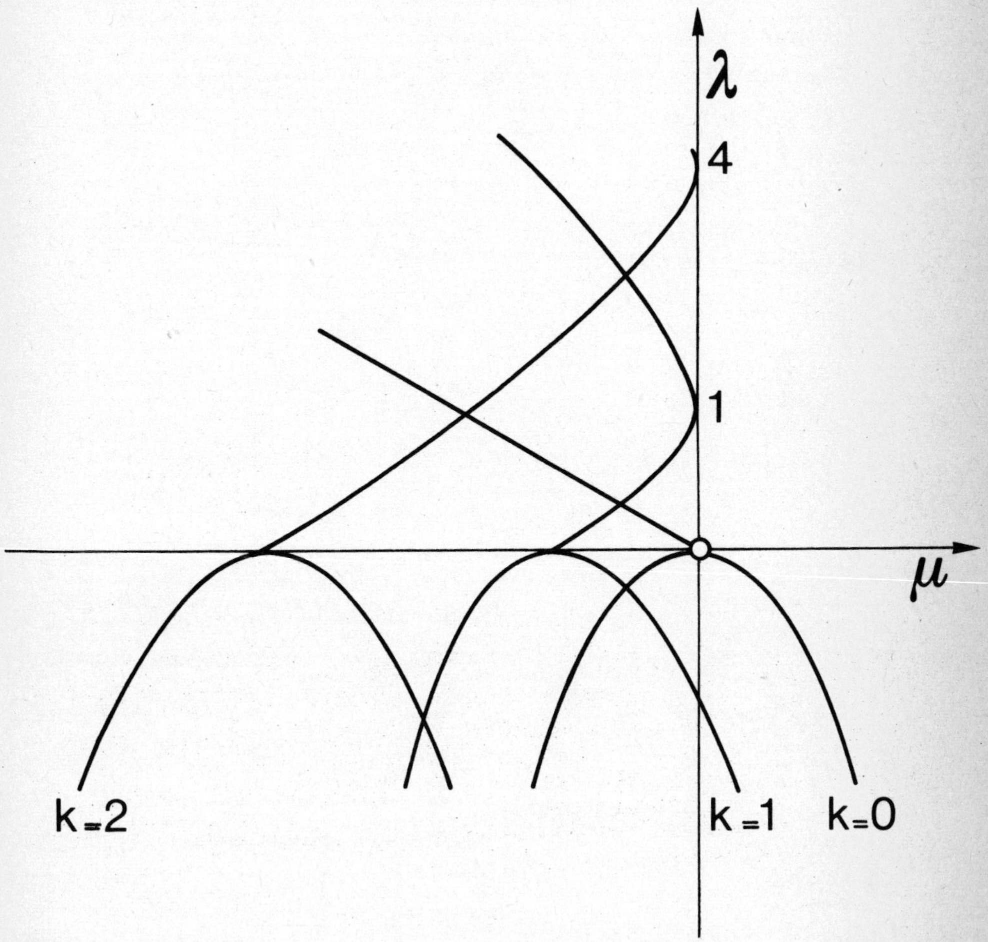

Figure 5

If one of the curves (3.12) is intersected transversally, then $\rho_r = 0$, $\rho_r' = 0$, and $\rho_i \neq 0$ if $\mu \neq 0$. Nonresonance is guaranteed if

(3.13) $$\tau \neq \frac{(k+1)^2}{4}$$

holds for all $(k,l) \in Z^2$, $k \neq \pm 1$. Hence Hopf-bifurcation occurs.

Proposition 3.7: For $\tau > 0$ and any point (μ,τ) on the curves given by (3.12), $(\tau,\mu,u=0)$ is a point of Hopf-bifurcation if (3.13) is satisfied.

An interesting situation arises when μ tends to 0 and thus ρ_r as well as ρ_i approach 0 simultaneously. This is the case which was treated in [2]. We show now, how it can be solved by the method developed here. Without loss of generality we can assume

$$\beta = -1 \quad , \qquad \alpha < 0 \quad , \quad \gamma \leq 0 \, ,$$

$$f_1 = u^3 \quad , \qquad f_2 = u^2 v \, .$$

Then we have for $\mu = \alpha + \gamma \to -0$

$$\tau = 1 - \frac{(\alpha-\gamma)^2}{4} > 0 \quad , \qquad \sigma_1 = \frac{\mu}{2} - i\sqrt{\tau} \, ,$$

$$\sigma_i = -\frac{1}{c}(\sigma_{1i} + 1) , \quad \sigma_2 = \frac{\mu}{2} + i\sqrt{\tau} \, .$$

The set S_o is given by $\{+1,-1\}$. We have to rewrite the equations for the Fourier-components (u^1,v^1) resp. (u^{-1},v^{-1}) in terms of the eigenvectors of A. If we denote these components by u_1, u_{-1}, we obtain to the lowest order

(3.14) $$u_1'' - cu_1' - (\sigma_1 + i)u_1 - g_1(u_1,u_{-1}) = 0$$

and for u_{-1} the conjugate complex equation

$$g_1 = \frac{1}{\sigma_1 - \sigma_2}((\alpha-\sigma_2)f_1 - f_2)^1 .$$

The superscript "1" indicates the Fourier-components for k = 1.

Of the two eigenvalues ρ corresponding to (3.14) only one is critical (the **other is** near c). Hence we can reduce by one dimension and obtain

(3.15) $$w_1' = \rho_{1\;1} - \frac{2}{c} w_1 |w_1|^2 \quad + \text{higher order terms} .$$

Setting $w_1 = R \exp(i\phi)$ one has to solve

$$R' = \rho_{1r} R - \frac{2}{c} R^3$$

$$\phi' = \rho_{1i}$$

Since

$$\rho_{1r} = - \frac{1}{c} \left(\frac{\mu}{2} + \frac{1}{c^2} (\sqrt{\tau} - 1)^2 \right)$$

holds, ρ_{1r} is positive if

$$- \frac{\mu}{2} > \frac{1}{c^2} (\sqrt{\tau} - 1)^2 .$$

This condition is satisfied in [2]. Set $R^* = (c\rho_{1r}/2)^{1/2}$, then the solution w_1 can be sketched in the complex plane

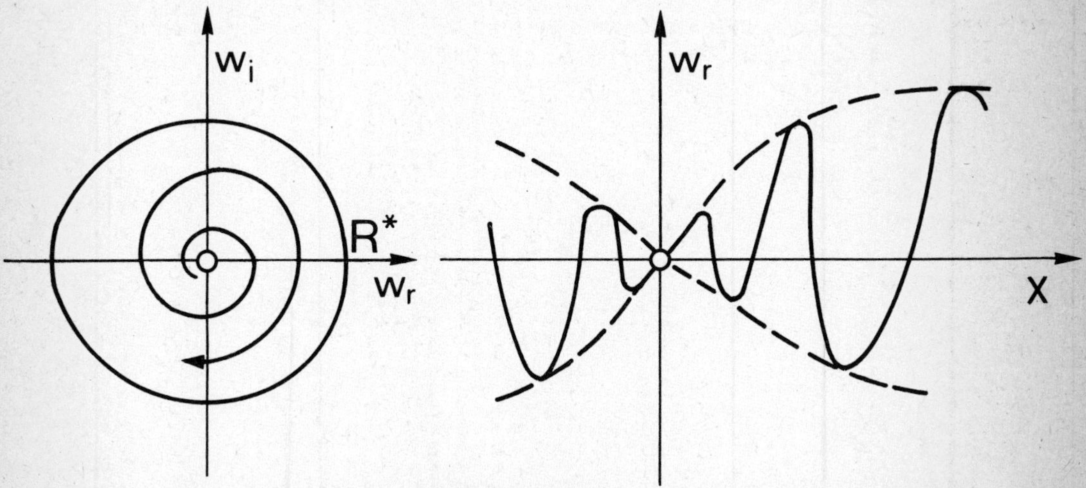

Figure 6

We can take real and imaginary part as real solutions and obtain R(x) $\cos \rho_{1i} x$.

References

[1] S.N. CHOW and J.K. HALE: Methods of bifurcation theory, Grundlehren der math. Wissenschaften Nr. 251, Springer-Verlag, 1982.

[2] D.S. COHEN, F.C. HOPPENSTEADT and R.M. MIURA: Slowly modulated oscillations in nonlinear diffusion processes, SIAM J. Appl. Math. 33 (1977), 217-229.

[3] P.C. FIFE: Mathematical aspects of reacting and diffusing systems, Lect. Notes in Biomathematics Nr. 28, Springer-Verlag, 1979.

[4] G. FISCHER:Zentrumsmannigfaltigkeiten bei elliptischen Differentialgleichungen, Math. Nachrichten 115 (1984), 137-157.

[5] D. HENRY: Geometric theory of semilinear parapolic equations, Lect. Notes in Math., Nr. 840, Springer-Verlag, 1981.

[6] K. KIRCHGÄSSNER: Wave-solutions of reversible systems, J. Diff. Equ. 45 (1982), 113-127.

[7] J. MOSER: Convergent series expansions for quasi-periodic motions, Math. Ann. 169 (1967), 136-176.

[8] J. SCHEURLE: Bifurcation of quasi-periodic solutions from equilibrium points of reversible dynamical systems, Arch. Rat. Mech. Anal., to appear.

VARIATIONAL METHODS AND BOUNDARY VALUE PROBLEMS FOR VECTOR SECOND ORDER DIFFERENTIAL EQUATIONS AND APPLICATIONS TO THE PENDULUM EQUATION

J. Mawhin and M. Willem

Université de Louvain
Institut Mathématique
B-1348 Louvain-la-Neuve
Belgique

Introduction

The first variational approach to the study of periodic solutions of the forced pendulum equation

$$u'' + a \sin u = e(t) \tag{1}$$

seems to be due to Hamel [3] in 1922. Using some results of the calculus of variation due to Hilbert and Caratheodory, he proved the existence of a solution of (1) with

$$e(t) = b \sin t \tag{2}$$

which satisfies the periodic boundary conditions

$$u(0) - u(2\pi) = u'(0) - u'(2\pi) = 0 \tag{3}$$

for arbitrary a and b.

When e is given by (2) or, more generally, when e is odd, Hamel also observed that 2π-periodic solutions of (1) could also be obtained by finding first a solution of (1) on $[0,\pi]$ satisfying the Dirichlet boundary conditions

$$u(0) = u(\pi) = 0, \tag{4}$$

and then extending it to $[-\pi,0]$ by the oddness relation

$$u(-t) = -u(t),$$

and then to R by 2π-periodicity. In contrast with (1)-(3), the boundary value problem (1)-(4) can easily be reduced to a nonlinear integral equation of the form

$$u(t) = -\int_0^{\pi} K(t,s)[a \sin u(s) - e(s)]ds. \tag{5}$$

Hammerstein [4] in 1930 used this equivalent form together with a combination of the Ritz and the variational method to prove the existence of a solution of (1)-(4) for arbitrary a and continuous e on $[0,\pi]$.

In this paper, we will review some results obtained by applying recent variational techniques to the problems of the above type, refering to the original papers [5-10] for further information.
This subject seems to us specially appropriate in a paper dedicated to Professor Cesari who, among so many other subjects, has given decisive contributions to the study of periodic solutions of differential equations and to various aspects of the calculus of variations and optimal control theory.

1. Variational methods and the Dirichlet problem for some second order systems of ordinary differential equations

Let $V : [0,b] \times R^n \to R$ be a function satisfying the Caratheodory conditions and such that $D_u V = (\frac{\partial V}{\partial u_1},\ldots,\frac{\partial V}{\partial u_n})$ exists on $[0,b] \times R^n$ and satisfies the Caratheodory conditions. Consider the Dirichlet boundary value problem, with $e \in L^1(0,b;R^n)$,

$$\begin{aligned} &u''(t) + D_u V(t,u(t)) = e(t), \\ &u(0) = u(b) = 0 . \end{aligned} \tag{6}$$

Let us denote by $H_0^1 = H_0^1(0,b;R^n)$ the Hilbert space of absolutely continuous functions $u : [0,b] \to R^n$ such that $u(0) = u(b) = 0$ and

$u' \in L^2(0,b;R^n)$, with the inner product

$$((u_1,u_2)) = \int_0^b (u_1'(t),u_2'(t))dt,$$

and the corresponding norm

$$\|u\| = \left(\int_0^b |u'(t)|^2 dt\right)^{\frac{1}{2}} = |u'|_{L^2},$$

where $(.,.)$ denotes the usual inner product in R^n.

Define the "action integral" on H_0^1 by

$$\phi(u) = \int_0^b \left[(1/2)|u'(t)|^2 - V(t,u(t)) + (e(t),u(t))\right]dt,$$

where $|.|$ denotes the Euclidian norm in R^n.

The following Lemma 1, which is very similar to Lemma 1 of [6], is basic for the proof of our existence result for (6).

Lemma 1. Let $\alpha \in L^1(0,b)$ be such that the inequality

$$\alpha(t) \leq \pi^2/b^2$$

holds a.e. on $[0,b]$, with the strict inequality on a subset of $[0,b]$ of positive measure. Then there exists $\delta = \delta(\alpha) > 0$ such that, for all $u \in H_0^1$, one has

$$(1/2)\int_0^b \left[|u'(t)|^2 - \alpha(t)|u(t)|^2\right]dt \geq \delta \|u\|^2 .$$

We can now state our first existence result and sketch its proof.

Theorem 1. Assume that there exists $\alpha = \alpha_0 + \alpha_1$ with $\alpha_i \in L^1(0,b)$ $(i=0,1)$ such that

(7) $$\limsup_{|u|\to\infty} |u|^{-2} V(t,u) \leq (1/2)\delta(t),$$

uniformly a.e. on $[0,b]$. Assume moreover that

$$\alpha_0(t) \leq \pi^2/b^2$$

a.e. on $[0,b]$, with the strict inequality on a subset of positive measure of $[0,b]$, and

$$|\alpha_1|_{L^1} < (6/b)\,\delta(\alpha_0) ,$$

where $\delta(\alpha_0)$ is associated to α_0 by Lemma 1.

Then, problem (6) has at least one classical solution minimizing ϕ on H_0^1. Moreover, if, for all $u \neq v$ in R^n, one has

$$(D_uV(t,u) - D_uV(t,v), u - v) < (\pi^2/b^2)\,|u-v|^2 , \tag{8}$$

then (6) has exactly one solution.

Sketch of the proof. By combining assumption (7) with Lemma 1, and using the well-known inequalities in H_0^1,

$$|u|_{L^\infty} \leq \sqrt{b/3}\,\|u\| , \quad |u|_{L^2} \leq (b/\pi)\,\|u\| ,$$

one can show the existence of $\gamma > 0$, $\varepsilon > 0$ and $\beta \in L^1(0,b)$ such that

$$\phi(u) \geq \gamma\|u\|^2 - \varepsilon|e|_{L^1}\|u\| - |\beta|_{L^1}$$

for all $u \in H_0^1$. ϕ being weakly lower semi-continuous, the existence of a minimum follows, which is a weak, and hence a classical solution of (6). The uniqueness under condition (7) follows from a classical argument.

Combining the result of Theorem 1 with the type of extension mentioned in the Introduction, one easily obtains existence results of odd 2b-periodic solutions for our problem under conditions which generalized the results of [10] to the case of non-uniform non-resonance conditions on the potential V. See [8] for more details.

Remark 1. By taking $n = 1$, $V(t,u) = - a \cos u$ and then $\alpha = 0$, we see that the problem

$$u'' + a \sin u = e(t),$$

$$u(0) = u(b) = 0$$

has, for each $a \in R$ and $e \in L^1(0,b)$ at least one solution minimizing the action integral ϕ given by

$$\phi(u) = \int_0^b [(1/2)u'^2(t) + a \cos u(t) - e(t)u(t)]\, dt .$$

Consider now the problem

$$(8) \qquad \begin{aligned} &u''(t) + D_u V(t,u(t)) = 0 , \\ &u(0) = u(b) = 0 \end{aligned}$$

and assume that

$$(9) \qquad D_u V(t,0) = 0$$

for a.e. $t \in [0,b]$. Consequently, (8) always admits the trivial solution $u = 0$ and we shall use the variational approach to find conditions insuring the existence of a non-trivial solution. See [10] for similar results for the existence of odd periodic solutions under uniform non-resonance conditions. Without loss of generality, we can assume that

$$(10) \qquad \int_0^b V(t,0)dt = 0.$$

Theorem 2. *Assume that* V *satisfies the conditions* (9) *and* (10) *and the the conditions of Theorem 1. If there exists* $c \in R^n$ *such that*

$$\int_0^1 V(bs,\ c \sin \pi s)ds > (\pi|c|)^2/4b^2, \tag{11}$$

then problem (8) has at least a non-trivial solution minimizing ϕ on H_0^1.

Proof. If u is the solution minimizing ϕ given by Theorem 1 and if

$$v(t) = (\sin(\pi/b)t)c\ ,$$

we get easily, using (11),

$$\phi(u) \leq \phi(v) = (\pi|c|)^2/4b - b \int_0^1 V(bs,\ c \sin\pi s)\ ds <$$
$$< 0 = \phi(0),$$

so that $u \neq 0$.

Remark 2. Consider the problem

$$\begin{aligned} &u'' + a(t) \sin u = 0, \\ &u(0) = u(b) = 0 \end{aligned} \tag{12}$$

with $a \in L^1(0,b)$, for which $V(t,u) = b^{-1} \int_0^b a(s)ds - a(t) \cos u$.

Now

$$\lim_{c \to 0} \int_0^1 c^{-2} V(bs, c \sin\pi s)ds = (1/2) \int_0^1 a(bs) \sin^2 \pi s\ ds\ ,$$

which implies that (12) will have a non-trivial solution if

$$\pi^2/2b^2 < \int_0^1 a(bs) \sin^2 \pi s\ ds\ . \tag{13}$$

In particular, for $a(t) = 1 + r \cos(\pi t/b)$, (13) becomes

$$b > \pi\ .$$

2. Variational methods and the periodic problem for some second order systems of ordinary differential equations

Let $V : [0,T] \times R^n \to R$ be a function satisfying the Caratheodory conditions and such that $D_u V$ exists on $[0,T] \times R$ and satisfies the Caratheodory conditions. Consider the periodic boundary value problem, with $e \in L^1(0,T;R^n)$,

$$u''(t) + D_u V(t,u(t)) = e(t) ,$$

(14)

$$u(0) - u(T) = u'(0) - u'(T) = 0 .$$

Let us denote by H the Hilbert space of absolutely continuous functions $u : [0,T] \to R^n$ such that $u(0) = u(T)$ and $u' \in L^2(0,T;R^n)$ with the inner product:

$$((u,v)) = \int_0^T \left[(u(t),v(t)) + (u'(t),v'(t)) \right] dt ,$$

and the corresponding norm $\|u\| = (|u|^2_{L^2} + |u'|^2_{L^2})^{1/2}$. If $u \in H$, we shall write $u = \bar{u} + \tilde{u}$, with

$$\bar{u} = T^{-1} \int_0^T u(t)dt , \quad \int_0^T \tilde{u}(t)dt = 0.$$

Define the "action integral" on H by

$$\phi(u) = \int_0^T \left[(1/2)|u'(t)|^2 - V(t,u(t)) + (e(t),u(t)) \right] dt.$$

The following result was first proved in [9] under slightly less general regularity assumptions.

<u>Theorem 3</u>. <u>Assume that there exist</u> $T_1 > 0, \ldots , T_n > 0$ <u>such that</u>

(15) $$V(t,u + T_i e_i) = V(t,u) \quad (1 \leq i \leq n)$$

<u>for a.e.</u> $t\in[0,T]$ <u>and all</u> $u\in R^n$, <u>with</u> $e_i = (\delta_{ij} : 1\le j\le n)$. <u>Then problem</u> (14) <u>has at least a solution</u> u <u>for every</u> $e\in L^1(0,T;R^n)$ <u>such that</u>

$$(16) \qquad \int_0^T e(t)dt = 0 ,$$

<u>and</u> u <u>minimizes</u> ϕ <u>on</u> H .

<u>Proof</u>. By (15), we have

$$(17) \qquad \left| \int_0^T V(t,u)dt \right| \le C$$

for all $u\in R^n$ and some $C \ge 0$. Now, using (16), (17) and the well-known inequalities

$$\|\tilde{u}\| \le \left(1 + \frac{T^2}{4\pi^2}\right)^{1/2} |\tilde{u}'|_{L^2} \quad , \quad |\tilde{u}|_{L^\infty} \le (1/2)(T/3)^{1/2}\|\tilde{u}\|$$

which hold for all $\tilde{u}\in H$ such that

$$\int_0^T \tilde{u}(t)dt = 0 ,$$

we obtain

$$\phi(u) = \int_0^T \left[(1/2)|\tilde{u}'(t)|^2 - V(t,\bar{u}+\tilde{u}(t)) + e(t)\tilde{u}(t) \right] dt \geq$$

$$(18) \qquad \geq \gamma \|\tilde{u}\|^2 - \delta |e|_{L^1} \|\tilde{u}\| - CT ,$$

for some $\gamma > 0$ and $\delta > 0$, so that $\phi(u) \to +\infty$ if $\|\tilde{u}\| \to \infty$. So, if (u_k) is a minimizing sequence for ϕ, then $(\tilde{u}_k)$ is bounded in H and, by (15) we can assume without loss of generality that

$$(u_k, e_i) \in [0,T_i] \quad (1\le i\le n),$$

so that (u_k) is bounded in H. The proof follows then in a classical way from the weak lower semi-continuity of ϕ.

Remark 3. For the problem

$$u'' + a \sin u = e(t),$$

$$(19) \qquad u(0) - u(T) = u'(0) - u'(T) = 0$$

we have $V(t,u) = -a \cos u = V(t,u+2\pi)$ and hence (18) has at least one solution for every $e \in L^1(0,T)$ such that $\int_0^T e(t)dt = 0$.

In the special case where $e = 0$, (19) has in fact the two distinct solutions $u = 0$ and $u = \pi$ having both mean value in $[0,2\pi]$. Such a situation still holds in the general case, as follows from Theorem 4 which is a slight extension of the result given in [7].

Theorem 4. Under the assumption of Theorem 3, the problem (14) has at least two solutions u_0 and v_0 such that $\|v_0 - u_0\| < T^{\frac{1}{2}} \min(T_1,\ldots,T_n)$ where

$$\phi(u_0) = \min_H \phi.$$

Sketch of the proof. The proof is based upon the Brézis-Coron-Nirenberg's modification [2] of the Ambrosetti-Rabinowitz mountain pass lemma [1]. Let $D\phi : H \to H^*$ denote the Gâteaux differential of ϕ, $\langle , \rangle$ the pairing between H and H^* and $J : H \to H^*$ the duality mapping. Let us denote by u_0 the solution given by Theorem 3, so that $\phi(u_0) = \min_H \phi$.

The application of the modified mountain pass lemma requires the verification of three conditions.

a. ϕ satisfies the weak Palais-Smale condition, i.e. if (u_k) in H is such that

$$(20) \qquad \phi(u_k) \to c \quad \text{and} \quad D\phi(u_k) \to 0,$$

then c is a critical value of ϕ. For all u and v in H, we have

$$(21) \qquad \langle D\phi(u),v\rangle = \langle Ju,v\rangle - \langle Nu,v\rangle$$

where $N : H \to H^*$ is defined by the relation

$$\langle Nu,v\rangle = \int_0^T (u(t) + D_u V(t,u(t)) - e(t),v(t))dt,$$

so that, by the compact embedding of H into $C([0,T],R^n)$, $Nu_k \overset{H^*}{\to} Nu$ when $u_k \overset{H}{\rightharpoonup} u$, where $\rightharpoonup$ denotes the weak convergence. Let (u_k) be a sequence such that (20) holds. By (15), we can assume that $(u_k,e_i) \in [0,T_i]$ $(1 \le i \le n)$ and, by (18), it follows that (u_k) is bounded, so that, up to a subsequence, we have $u_k \overset{H}{\rightharpoonup} u$ for some $u \in H$. By (21),

$$Ju_k \overset{H^*}{\to} Nu ,$$

and hence $u_k \overset{H}{\to} u$, $D\phi(u) = 0$ and $\phi(u) = c$.

b. <u>There exists</u> $R > 0$ <u>and</u> $\rho > (u_0)$ <u>such that</u> $\phi(u) > \rho$ if $\|u-u_0\| = R$.

Notice first that if u_0 is not a strict local minimum for ϕ, then our Theorem 4 is proved. If u_0 is a strict local minimum, then there exists $R > 0$ such that $\phi(u) > \phi(u_0)$ when $0 < \|u - u_0\| \leqslant R$. From (15) we have necessarily $R < T^{1/2} \min(T_1,\dots,T_n)$. Let

$$\rho = \inf_{\|u-u_0\| = R} \phi(u) \ge \phi(u_0).$$

If $\rho > \phi(u_0)$ we are done. If $\rho = \phi(u_0)$, there is a sequence (u_k) such that $u_k \overset{H}{\rightharpoonup} u$, $\phi(u_k) \to \phi(u_0)$ and $\|u-u_0\| \le R$; if $u \ne u_0$, then $\phi(u) > \phi(u_0) = \lim_{k\to\infty} \phi(u_k) \ge \phi(u)$, a contradiction.

If $u = u_0$, as $\phi(v) = (1/2)\|v\|^2 + \psi(v)$ with ψ continuous for the uniform topology, and as H is compactly imbedded in $C([0,T];R^n)$, we can assume, up to a subsequence, that $u_k \overset{C}{\to} u_0$, so that $(1/2)\|u_k\|^2 = \phi(u_k) - \psi(u_k) \to \phi(u_0) - \psi(u_0) = (1/2)\|u_0\|^2$, and $u_k \to u_0$, a contradiction with $\|u_k - u_0\| = R$.

c. <u>There exists</u> $v \in H$ <u>such that</u> $\|v-u_0\| > R$ <u>and</u> $\phi(v) < \rho$.

Taking, say, $v = u_0 + T_1 e_1$, we have $\phi(v) = \phi(u_0) < \rho$, and

$$\|v - u_0\| = T^{1/2} T_1 > R.$$

Remark 4. When e has mean value zero, the problem (19) has at least two solutions which do not differ by a multiple of 2π. On the other hand, a necessary condition for (19) having a solution is that

$$\bar{e} \in [-a,a].$$

If we write $e = \bar{e} + \tilde{e}$, it follows from results of [7] , that we shall not develope here, that, for each $\tilde{e} \in C([0,T];R^n)$ having mean value zero, the set of $\bar{e} \in R$ such that (19) with $e = \bar{e} + \tilde{e}$ has a solution is a closed interval $[d(\tilde{e}), D(\tilde{e})]$ containing 0 and that the set of $\tilde{e} \in C([0,T]; R^n)$ with mean value zero for which $d(\tilde{e}).D(\tilde{e}) < 0$ is open and dense in that space.

Bibliography

[1] A. AMBROSETTI, P. RABINOWITZ - Dual variational methods in critical point theory and applications, J. Functional Anal. 14 (1973) 349-381.

[2] H. BREZIS, J.M. CORON, L. NIRENBERG - Free vibrations for a nonlinear wave equation and a theorem of P. Rabinowitz, Comm. Pure Appl. Math. 33 (1980) 667-684.

[3] G. HAMEL - Uber erzwungene Schwingungen bei endlichen Amplituden, Math. Ann. 86 (1922) 1-13.

[4] A.HAMMERSTEIN, Nichtlineare Integralgleichungen nebst Anwendungen, Acta Math. 54 (1930) 117-176.

[5] J. MAWHIN - Periodic oscillations of forced pendulum-like equations, in "Proceedings Conference on Differential Equations", Dundee, 1982 , Lect. Notes Math. n. 964, Springer, Berlin, 1982, 458-476.

[6] J. MAWHIN, J. WARD Jr. - Periodic solutions of some forced Liénard differential equations at resonance, Arch.Math.(Basel) 41 (1983) 337-351.

[7] J. MAWHIN, M. WILLEM - Multiple solutions of the periodic boundary value problem for some forced pendulum-type equations, J. Differential Equations, 52 (1984) 264-287.

[8] J. MAWHIN, J. WARD Jr., M. WILLEM, Variational methods and semilinear elliptic equations, Arch. Rat. Mech. Anal., to appear.

[9] M. WILLEM - Oscillations forcées de systèmes hamiltoniens, Public. Séminaire Analyse non linéaire de l'Univ. de Besançon, 1981.

[10] M. WILLEM - Periodic oscillations of odd second order Hamiltonian systems, Bol. Un. Mat. Ital., in press.

STABILITE DE REGIME DES MACHINES TOURNANTES ET PROBLEMES ASSOCIES

M. ROSEAU

Laboratoire de Mécanique Théorique
associé au C.N.R.S.
Tour 66, 4 Place Jussieu - 75230 - Paris

Les systèmes vibrants dont on se propose de discuter la stabilité dans ce texte peuvent être définis, au regard de leur configuration à un instant donné, par une variable angulaire vectorielle, associée aux rotations d'arbres de la machine et notée $\phi = \{\phi_1, \phi_2, \ldots, \phi_p\}$, à valeurs sur le tore T^p, et une ou plusieurs variables vectorielles x,y aptes à décrire le mouvement de la structure portante.

Dans le régime idéal, les vitesse angulaires $\frac{d\phi_j}{dt}$ prennent des valeurs constantes ω_j et la structure est en repos ce qui implique x,y constants. Mais pour des raisons très diverses, équilibrage dynamique imparfait, couplage entre la source d'énergie et la machine, flexibilité anisotrope des paliers, anisotropie structurelle des arbres, flexion et mouvement tourbillonnaire de ceux-ci, etc, des effets d'excitation, souvent très complexes entrent en jeu qui, éventuellement, peuvent être à l'origine d'instabilité du régime nominal de fonctionnement.

Pour en faire l'étude on est ainsi conduit à écrire les équations différentielles du mouvement, prenant en compte les effets dont on pense qu'ils peuvent jouer un rôle essentiel dans l'apparition d'instabilité et, dans cette voie, on peut proposer des équations modèles qui seront l'objet d'investigation théorique. Plus loin quelques exemples seront présentés pour illustration.

Systemes multiperiodiques

On peut considérer successivement les trois systèmes suivants dans l'ordre de complexité croissante:

(1) $\frac{d\phi}{dt} = \omega + h(\phi,\mu)$ avec $\omega = \{\omega_1,\ldots,\omega_p\} \in \mathbb{R}^p$

$$(k,\omega) = \sum_1^p k_j\omega_j \neq 0, \quad \forall k \in Z^p - 0$$

(on dira que les pulsations ω_j sont indépendantes)

$h(\phi,\mu)$ à valeurs dans $\mathbb{R}^p$, périodique par rapport à chaque ϕ_j, de période de 2π, ou encore définie sur le tore T^p, dépendant aussi d'un petit paramètre μ,ω et h donnés.

(2)
$$\frac{dx}{dt} = Ax+F(x,\phi,\mu)$$
$$\frac{d\phi}{dt} = \omega+h(x,\phi,\mu)$$

A matrice constante $\{n \times n\}$ stable, c'est-à-dire à valeurs propres de partie réelle négative.
F et h à valeurs dans $\mathbb{R}^n$, $\mathbb{R}^p$ respectivement, multipériodiques en ϕ.

(3)
$$\frac{dx}{dt} = Ax+F(x,y,\phi,\mu,)$$
$$\frac{dy}{dt} = By+f(x,y,\phi,\mu)$$
$$\frac{d\phi}{dt} = \omega+h(x,y,\phi,\mu)$$

A matrice constante $\{n \times n\}$ stable, B matrice constante $\{q \times q\}$ critique, c'est-à-dire à valeurs propres imaginaires pures $i\,\lambda_j$, λ_j réel, $1 \leq j \leq q$.

On sait que la recherche pour ces systèmes de solution quasi périodique se heurte à la difficulté des petits diviseurs qui est que $|(\omega,k)^{-1}|$ peut devenir très grand par un choix convenable de $k \in Z^p - 0$. Cependant on dispose d'une information très utile grâce au lemme suivant [1]: il existe une constante positive $c(\omega)$ telle que pour presque tout ω dans le cube $C = \{\omega \in \mathbb{R}^p,\ 0 < \omega_j < 1,\ 1 \leq j \leq p\}$, on ait:

$$|(\omega,k)|^{-1} < c|k|^{p+1} \quad , \quad |k| = \sum_{j=1}^{p} |k_j|, \quad \forall k \in Z^p - 0.$$

Grâce à la méthode de convergence accélérée introduite par Kolmogoroff, l'existence de solution quasi périodique pour les systèmes du type (1) et (2) a pu être établie lorsque les pulsations ω obéissent à la condition du lemme; celle-ci, certes, n'est pas très contraignante d'un point de vue statistique mais il est cependant difficile de savoir, dans une situation donnée si elle est remplie ou non, faute de pouvoir caractériser simplement dans l'espace ω l'ensemble d'exclusion de mesure nulle.

Pour le système (3) cette condition doit d'ailleurs être complétée ainsi qu'il suit:

le couple ω,λ doit être tel qu'il existe une constante numérique c et un entier τ positifs tels que:

$$|\lambda_r - (k,\omega)|^{-1} < c|k|^{\tau} \qquad \forall\ r,s \in [1,2,\dots,q] \times [1,\dots,q],$$

$$|\lambda_r - \lambda_s - (k,\omega)|^{-1} < c|k|^{\tau} \qquad \forall\ k \in Z^p - 0.$$

Cette situation peut encore relever du lemme cité en considérant le point pulsation ω,λ dans un espace de dimension $p + q$, et appelle les mêmes commentaires.

Supposant que μ est un petit paramètre réel positif et sous certaines hypothèses complémentaires on peut formuler les résultats [2,3] qui suivent:

1) avec $h(\phi,\mu)$ analytique en ϕ dans $|\mathrm{Im}\ \phi| < b$, multipériodique en ϕ, et $|h| < \mu$ dans cette bande, il existe $\varepsilon(\mu) \in R^p$, $\lim_{\mu\to 0} \varepsilon(\mu) = 0$ et $\phi(\Theta,\mu)$ analytique en $\Theta = (\Theta_1,\dots,\Theta_p)$ dans $|\mathrm{Im}\ \Theta| < b/2$, $\phi(\Theta,\mu) - \Theta$ multipériodique, satisfaisant avec $\Theta = \omega t$ et μ assez petit à l'equation:

$$\frac{d\phi}{dt} = \omega + \varepsilon(\mu) + h(\phi,\mu) \ ;$$

2) avec $h(x,\phi,\mu)$ analytique en x,y dans $||x|| < R$, $|\text{Im } \phi| < b$, multipériodique en ϕ et satisfaisant dans cet ouvert à:

$$||F|| < a_1\mu + a_2 ||x||^2 \quad , \quad ||h|| < b_1\mu + b_2 ||x||$$

où a_j, b_j sont des constantes, il existe $\varepsilon(\mu) \in \mathbb{R}^p$, $\lim_{\mu\to 0} \varepsilon(\mu) = 0$ et $x(\Theta,\mu)$, $\phi(\Theta,\mu)$ analytiques en $\Theta = \{\Theta_1,\ldots,\Theta_p\}$ dans $|\text{Im } \Theta| < b/2$, $x(\Theta,\mu)$, $\phi(\Theta,\mu)-\Theta$ multipériodiques et satisfaisant avec $\Theta = \omega t$ et μ assez petit à:

$$\frac{dx}{dt} = Ax + F(x,\phi,\mu)$$

$$\frac{d\phi}{dt} = \omega + \varepsilon(\mu) + h(x,\phi,\mu) ;$$

3) l'analyse du système (3) montre que sous certaines hypothèses concernant F, f, h (analyticité en x,y,ϕ, multipériodicité, estimations convenables [4]), on peut établir l'existence d'éléments $Y(\mu), m(\mu), \varepsilon(\mu)$ respectivement matrice et vecteurs, tous tendant vers 0 avec μ, tels que le système modifié:

$$\frac{dx}{dt} = Ax + F(x,y,\phi,\mu)$$

$$\frac{dy}{dt} = (B+Y)y + m + f(x,y,\phi,\mu)$$

$$\frac{d\phi}{dt} = (\omega+\varepsilon) + h(x,y,\phi,\mu)$$

ait une solution $x(\Theta,\mu)$ $y(\Theta,\mu)$, $\phi(\Theta,\mu)$ avec $x(\Theta,\mu)$, $y(\Theta,\mu)$ $\phi(\Theta,\mu)-\Theta$ multipériodique en Θ de période 2π et $\Theta = \omega t$.

Ainsi dans les trois cas considérés, l'existence d'une solution quasi périodique de pulsation ω_j est subordonnée d'une part, à des conditions de nature arithmétique sur les pulsations, très difficiles à reconnaître concrètement et, d'autre part, à la nécessité de modifier le second membre des équations par l'addition de termes ayant une structure simple connue (constante ou affine), mais dont il est impossible en général d'apprécier numériquement les coefficients; c'est pourquoi, même si dans certains cas, grâce à la présence dans les équations du mouvement de certaines constantes provenant d'intégrations antérieures, il apparaît théoriquement possible de réaliser les ajustements qui conduisent aux équations modifiées, ceux-ci ne pourront être réalisés numériquement.

Les difficultés qui sont sous jacentes à l'application de ces résultats pourtant très profonds conduisent à se demander si, moins exigeant quant à la nature de la solution recherchée, on n'en pourrait pas moins obtenir des résultats d'utilité pratique basés sur des hypothèses beaucoup moins contraignantes.

Dans cette voie, on peut formuler le résultat suivant |5| pour le système (2):

si A est stable, F et h continument différentiables en x, et $\|F\| < a_1\mu + a_2\|x\|^2$, $\|h\| < b_1\mu + b_2\|x\|$ dans $\|x\| < R$, $(\omega,k) \neq 0$ $\forall\, k \in Z^p - 0$, F et h multipériodiques en ϕ, toutes les variables réelles, alors pour μ et $\|x(0)\|$ assez petits, toute solution x,ϕ issue du point $x(0)$, $\phi(0)$ est définie dans l'avenir et telle que:

$$\lim_{\substack{x(o)\to o \\ \mu\to o}} \ \sup_{t\geq o} \left(\|x(t)\| + \left\|\frac{d\phi}{dt} - \omega\right\|\right) = 0 .$$

Un résultat analogue est valable pour :

$$\frac{dx}{dt} = \mu Ax + \mu F(x,\phi,\mu)$$

$$\frac{d\phi}{dt} = \omega + h(x,\phi,\mu)$$

auquel peuvent être réduits des systèmes du type:

$$\frac{da}{dt} = \mu A(a,\Theta) + \mu\tilde{A}(a,\Theta,\mu)$$

$$\frac{d\Theta}{dt} = \omega + \mu\ B(a,\Theta) + \mu\ \tilde{B}(a,\Theta,\mu)$$

A, B, $\tilde{A}$, $\tilde{B}$ fonctions continument différentiables de $(a,\Theta) \in \bar{G} \times T^p$, G ouvert de $\mathbb{R}^n$, et

$$\lim_{\mu\to 0} \ \sup_{(a,\Theta)\in \bar{G} \times T^p} (\|\tilde{A}\| + \|\tilde{B}\|) = 0$$

systèmes qu'on rencontre naturellement dans de nombreux systèmes mécaniques vibrants (oscillateurs couplés par des effets non linéaires, avec ou sans résonance).

Il est fréquent que l'analyse de stabilité conduise à s'interroger sur l'évolution quand $t \to +\infty$, pour $\mu > 0$ assez petit, de solution du système linéaire:

$$(4) \qquad \frac{da}{dt} = \mu[A(\Theta) + \mu A_1(\Theta) + O(\mu^2)]\ a \qquad \frac{d\Theta}{dt} = \omega$$

avec $\omega = \{\omega_1,\ldots,\omega_p\}$ système donné de pulsations indépendantes et $A(\Theta)$, $A_1(\Theta)$ matrices $\{n \times n\}$ dont les éléments sont des fonctions polynômes de $e^{i\Theta_j}$, $1 \leq j \leq p$, $O(\mu^2)$ terme dépendant de Θ mais uniformément d'ordre 2 en μ.

Avec $P(\Theta)$ matrice polynôme définie plus loin, on peut calculer les matrices polynômes $U(\Theta)$, $U_1(\Theta)$ et les matrices constantes S,S_1 telles que:

$$\sum_{j=1}^{p} \omega_j \frac{\partial U}{\partial \Theta_j} = A(\Theta) - S \quad , \quad \sum_{j=1}^{p} \omega_j \frac{\partial U_1}{\partial \Theta_j} = P(\Theta) - S_1$$

et faisant le changement de variable

$$a = (I+\mu U+\mu^2 U_1)\xi$$

il vient après un calcul simple:

$$\frac{d\xi}{dt} = \mu(S+\mu(AU-US+A_1-P+S_1)+O(\mu^2))\xi.$$

Choisissant $P = A U + A_1 - U S$, matrice polynôme, il reste:

$$\frac{d\xi}{dt} = \mu(S+\mu S_1+O(\mu^2))\xi .$$

Si les valeurs propres de S sont toutes de partie réelle négative, il y a stabilité de la solution $a = 0$; mais si certaines d'entre elles sont imaginaires pures, le recours au terme du second ordre est nécessaire et la stabilité peut être discutée d'après les propriétés de la matrice constante $S + \mu S_1$; si l'indétermination subsiste à ce stade, on pourra envisager de construire une approximation du troisième ordre, suivant un processus analogue. Le calcul des zones d'instabilité d'un rotor anisotrope sur suspensions élastiques anisotropes sera basé plus loin sur l'emploi de cette méthode.

Systèmes periodiques

On considère le cas où l'on peut décrire le mouvement par le moyen d'une variable angulaire scalaire ϕ et de variables vectorielles de structure x,y. Plusieurs systèmes standard peuvent être considérés.

a) $$\frac{d^2\phi}{dt^2} = \mu f(x,y,\frac{d\phi}{dt},\phi,\mu)$$

(5) $$\frac{dx}{dt} = Ax+h(y,\frac{d\phi}{dt},\phi,\mu)$$

$$\frac{dy}{dt} = \mu g(x,y,\frac{d\phi}{dt},\phi,\mu)$$

avec f, g, h périodiques en ϕ de période 2π, A matrice stable, μ petit paramètre, constitue un système autonome, d'ailleurs fortement non linéaire dont on trouvera la résolution et une application à un mécanisme de régulateur dans [6].

b)

(6)

$$\frac{d\phi}{dt} = \omega_1 + \mu\omega$$

$$\mu \frac{d\omega}{d\phi} = a(\phi)\omega + h(\phi) + \mu g(x,\omega,\phi)$$

$$\frac{dx}{d\phi} = Bx + k(\omega,\phi) + \mu l(x,\omega,\phi)$$

avec a, h, g, k, l périodiques en ϕ de période 2π, ω_1 constant, B matrice constante critique, telle que $\frac{dx}{d\phi} = Bx$ n'ait pas de solution périodique de période 2π, est un système à perturbation singulière pour μ petit, qu'on discutera plus loin sur le modèle d'un arbre imparfaitement équilibré sur paliers élastiques et soumis à un couple moteur de caractéristique raide.

c) Le système non autonome

(7)

$$\frac{d^2\phi}{dt^2} = \mu f(y,\phi,\frac{d\phi}{dt},t,\phi)$$

$$\frac{dy}{dt} = \mu g(y,\phi,\frac{d\phi}{dt},t,\phi)$$

avec f, g périodiques en ϕ de période 2π, périodiques en t de période T, apparaît dans plusieurs mécanismes de synchronisation et son traitement peut être réduit à celui de:

(8)

$$\frac{dx}{dt} = \upsilon f(x,y,t,\upsilon) \quad , \quad \frac{dy}{dt} = \upsilon^2 g(x,y,t,\upsilon)$$

où x, y sont des variables vectorielles, f et g périodiques en t de période T (de signification autre que dans (7)), υ un petit paramètre. Si f et g sont continus et de classe C^2 par rapport aux variables x, y, υ dans le voisinage du point $(x_o, y_o, 0)$ défini plus loin, on peut établir pour (8) un résultat d'existence de solution périodique et discuter les conditions de stabilité de celle-ci sous les hypothèse suivantes:

soit x_o, y_o un point de synchronisation défini par

$$\int_0^T f(x_o,y_o,t,o)\,dt=0 \quad , \quad \int_0^T g(x_o,y_o,t,o)\,dt=0$$

et $P_{ij}(t)$ les matrices périodiques en t:

$$P_{11}(t)=f_x(x_o,y_o,t,o) \quad , \quad P_{12}(t)=f_y(x_o,y_o,t,o) \quad ,$$

$$P_{21}(t) = g_x(x_o,y_o,t,o) \quad , \quad P_{22}(t)=g_y(x_o,y_o,t,o)$$

où f_x, f_y, g_x, g_y désignent les matrices des dérivées partielles.

Avec $S_{ij} = \frac{1}{T} \int_0^T P_{ij}(t)dt$, si la matrice $\begin{pmatrix} S_{11} & S_{12} \\ S_{21} & S_{22} \end{pmatrix}$ est inversible, le système (8) possède pour υ assez petit une solution périodique de période T qui tend vers x_o, y_o quand $\upsilon \to 0$.

La discussion de la stabilité requiert en général qu'on ait de cette solution une représentation asymptotique au deuxième ordre qu'on peut obtenir comme suit: soit ζ_o , χ_o la solution unique du système linéaire:

$$S_{11}\zeta_o+S_{12}\chi_o + \overline{P_{11}(t)\int_0^t f(x_o,y_o,t,o)\,dt} + \bar{f}_\upsilon = 0$$

$$S_{21}\zeta_o+S_{22}\chi_o + \overline{P_{21}(t)\int_o^t f(x_o,y_o,t,o)dt+g_\upsilon} = 0$$

avec $f_\upsilon = \frac{\delta f}{\delta \upsilon}$, $g_\upsilon = \frac{\delta g}{\delta \upsilon}$, la barre signifiant l'opérateur de moyenne sur une période T.

Alors on peut représenter la solution périodique voisine du point de synchronisation x_o, y_o par:

$$x(t,\upsilon) = x_o+\upsilon(\zeta_o+ \int_o^t f(x_o,y_o,t,o)dt) + O(\upsilon^2),$$

$$y(t,\upsilon) = y_o+\upsilon\chi_o + O(\upsilon^2)$$

qui permet le calcul des matrices $Q_{ij}(t)$ qui apparaissent au premier ordre en υ dans le développement de:

$$f_x(x(t,\upsilon),y(t,\upsilon),t,\upsilon)=P_{11}(t)+\upsilon Q_{11}(t)+O(\upsilon^2)$$

$$f_y(x(t,\upsilon),y(t,\upsilon),t,\upsilon)=P_{12}(t)+\upsilon Q_{12}(t)+O(\upsilon^2).$$

On peut ensuite définir les matrices périodiques $U_{ij}(t)$ de période T, par $\frac{d\,U_{ij}}{d\,t} = P_{ij}(t)-S_{ij}$ et avec:

$$R_{11} = S_{11}\,U_{11} - U_{11}\,P_{11} + Q_{11}$$

$$R_{12} = S_{11}\,U_{12} - U_{11}\,P_{12} + Q_{12}$$

on calcule

$$\Sigma_{1j} = \frac{1}{T}\int_o^T R_{1j}(t)dt \quad , \quad j = 1,2.$$

La stabilité de la solution peut être discutée à partir des propriétés de la matrice $\begin{pmatrix} S_{11}+\upsilon\,\Sigma_{11} & S_{12}+\upsilon\Sigma_{12} \\ \upsilon\,S_{21} & \upsilon\,S_{22} \end{pmatrix}$.

De manière précise il y aura stabilité pour $\upsilon > 0$ assez petit si sont remplies les conditions suivantes:

a) les racines de l'équation dét $\begin{pmatrix} S_{11} & S_{12} \\ S_{21} & S_{22-\sigma} \end{pmatrix} = 0$

sont simples et de partie réelle négative;

b) les racines de l'équation dét $(S_{11}-\lambda) = 0$ sont à partie réelle négative ou sont imaginaires pures non nulles. Ces dernières, si elles existent, sont simples et pour chacune d'elles λ_o le nombre

$$\sigma = - \left. \frac{\Theta_\upsilon}{\Theta_\lambda} \right/ \begin{array}{l} \upsilon = 0 \\ \lambda = \lambda_o \end{array} ,$$

avec $\Theta(\lambda,\upsilon) = \text{dét} \begin{pmatrix} S_{11} + \upsilon\Sigma_{11} - \lambda I & S_{12} + \upsilon\Sigma_{12} \\ \upsilon S_{21} & \upsilon S_{22} - \lambda I \end{pmatrix}$

est de partie réelle négative.

Ce ne sont là que conditions suffisantes qu'on pourra élargir dans certains cas.

Instabilité d'une machine tournante sur paliers elastiques avec couple moteur a caractéristique raide

L'arbre supposé rigide et horizontal repose sur deux paliers assurant une réponse élastique dans la direction verticale; le mouvement est supposé décrit par deux paramètres, ϕ angle de rotation de l'arbre et y qui mesure le déplacement vertical. D'autre part l'arbre est supposé imparfaitement équilibré et l'on admet que le moment du couple moteur dépend de la vitesse de rotation suivant une loi $M = M_o - h\,\phi'$, M_o et h constants.

Supposant en outre que le moment du couple résistant peut être représenté par $k\,\phi'$, $k > 0$ constant, notant m_1 la masse du balourd qui représente le défaut d'équilibrage, m_2 la masse de la machine et de l'arbre en mouvement vertical, $m=m_1+m_2$, J_1 le moment d'inertie de l'arbre et $J=J_1+m_1r^2$ avec r distance du balourd à l'axe de l'arbre c la rigidité de la suspension élastique, les équations du mouvement sont:

$$(9)\quad \begin{aligned} & my''-m_1r(\phi''\sin\phi+\phi'^2\cos\phi)+cy = 0 \\ & J\phi''-m_1ry''\sin\phi = M_o-(h+k)\phi'. \end{aligned}$$

Un étude expérimentale de ce problème conduite il y a longtemps par Sommerfeld fait apparaître une zone d'instabilité pour la vitesse de rotation; le problème a été discuté sur une base théorique par plusieurs auteurs [7,8] mais la méthode utilisée, de caractère heuristique, est très contestable et les résultats qu'elle fournit sont incorrects. D'ailleurs ces auteurs n'ont pas reconnu le caractère de perturbation singulière du système différentiel sur la base duquel peut être édifiée une théorie correcte [9]. En effet on admet que la caractéristique du moteur d'entrainement est raide c'est-à-dire qu'il faut une forte variation du couple moteur M pour réaliser une petite variation de la vitesse de rotation ϕ'.

Les équations réduites obtenues à partir de (9), complétées par l'introduction d'un terme d'amortissement pour la suspension, peuvent s'écrire:

$$(10)\quad \begin{aligned} & y''-ar\phi'^2\sin\phi = -\omega_o^2\, y-\mu f\omega_o y'+ar\phi'^2\cos\phi \\ & \phi''-br^{-1}y''\sin\phi = \mu^{-1}\omega_o(\omega_1-\phi') \end{aligned}$$

où $\quad a = \dfrac{m_1}{m_1+m_2} < 1 \quad , \quad b = \dfrac{m_1r^2}{J_1+m_1r^2} < 1 \quad , \quad \omega_o^2 = \dfrac{c}{m}$,

$$(11)\quad \omega_1 = \frac{M_o}{h+k} \quad , \quad \mu = \frac{J\omega_o}{h+k}$$

et l'hypothèse de caractéristique raide signifie que μ est un petit paramètre.

Avec $z=y'$, $\phi' = \omega_1 + \mu\omega$, on introduit y, z, ω pour nouvelles variables d'état du système, fonction de ϕ d'ailleurs lié au temps par $t = \int \frac{d\phi}{\omega_1+\mu\omega(\phi)}$; utilisant aussi $\tilde{\omega} = \omega + b\ \omega_o r^{-1} y \sin\phi$, on peut dès lors récrire (10) sous la forme:

$$\mu \frac{d\tilde{\omega}}{d\phi} = \frac{ab\omega_1^2 \sin\phi\cos\phi-\omega_o\tilde{\omega}}{\omega_1(1-ab\sin^2\phi)}$$

(12)

$$-\mu\Big[br^{-1}\frac{\omega_o}{\omega_1} z(f-1+ab\sin^2\phi)\sin\phi+b\omega_o r^{-1}y(-1+2ab\sin^2\phi)\cos\phi$$

$$+b\frac{\omega_o^2 r^{-1}}{\omega_1^2} y\sin\phi\cdot\tilde{\omega}-\frac{\tilde{\omega}}{\omega_1^2}(ab\omega_1^2\sin\phi\cos\phi+\omega_o\tilde{\omega})\Big]\ (1-ab\sin^2\phi)^{-1}+O(\mu^2)$$

$$\frac{dy}{d\phi} = \omega_1^{-1}\ z-\mu\Big(\frac{\tilde{\omega}}{\omega_1^2}z-\frac{b\omega_o r^{-1}}{\omega_1^2}\ \sin\phi\cdot yz\Big)+O(\mu^2)$$

$$\frac{dz}{d\phi} = -\frac{\omega_o^2}{\omega_1} y+\frac{ar\omega_1^2\cos\phi-ar\omega_o\tilde{\omega}\sin\phi}{\omega_1(1-ab\sin^2\phi)}$$

$$+\mu\Big[\omega\frac{\omega_o^2}{\omega_1^2}y+(1-ab\sin^2\phi)^{-1}(2ar\omega\cos\phi-f\frac{\omega_o}{\omega_1}z-\frac{\omega}{\omega_1^2}(\omega_1^2\cos\phi-\omega_o\tilde{\omega}\sin\phi)ar)\Big]$$

$$+O(\mu^2).$$

Ainsi il apparaît que le système (12) peut être discuté dans le cadre d'une théorie générale pour

(13)

$$\mu\frac{dx}{d\phi} = A(\phi)x+h(\phi)+\mu g(x,y,\phi,\mu)$$

$$\frac{dy}{d\phi} = By+k(x,\phi)\ +\ \mu h(x,y,\phi,\mu)$$

où x, y sont des variables vectorielles, h, g, k, ℓ sont périodiques en ϕ de période 2π, $A(\phi)$ matrice à valeurs propres de partie réelle moindre qu'une constante négative -2β et B une matrice constante à valeurs propres imaginaires pures, telle cependant que $\frac{dy}{d\phi} = By$ n'ait pas de solution périodique de période 2π, ce qui dans le cas particulier s'exprime par $\frac{\omega_0}{\omega_1}$ non entier. Transposant (13) à (12) on voit que x, y seraient de dimension 1 et 2 et $k=q(\phi)\ x+c(\phi)$. On peut établir au sujet de (13) qu'il existe pour μ assez petit une solution périodique $x(\phi,\mu)$, $y(\phi,\mu)$ de période 2π, qui, lorsque $\mu\to 0$, tend vers la solution périodique $\xi(\phi)$, $\zeta(\phi)$ unique définie par

$$A(\phi)\xi(\phi)+h(\phi) = 0 \quad , \quad \frac{d\zeta}{d\phi} = B\zeta + k(\xi(\phi),\phi).$$

Pour discuter la stabilité de cette solution il convient d'analyser le comportement des solutions du système linéaire aux variations c'est-à dire:

$$\mu \frac{du}{d\phi} = A(\phi)\ u + \mu P(\phi,\mu)u+\mu\ Q(\phi,\mu)\ v$$

$$\frac{dv}{d\phi} = Bv + H(\phi,\mu)u + \mu\ R(\phi,\mu)\ v$$

avec

$$P(\phi,\mu)=g_x(x(\phi,\mu),y(\phi,\mu),\phi,\mu) \ ,$$

$$Q(\phi,\mu)=g_y(x(\phi,\mu),y(\phi,\mu),\phi,\mu) \ ,$$

$$R(\phi,\mu)=h_y(x(\phi,\mu),y(\phi,\mu)\phi,\mu) \ ,$$

$$H(\phi,\mu)=k_x(x(\phi,\mu),\phi)+\mu l_x(x(\phi,\mu),y(\phi,\mu),\phi,\mu).$$

On calcule aisément les limites $P(\phi),\ldots, H(\phi)$ de $P(\phi,\mu),\ldots,H(\phi,\mu)$ quand $\mu\to 0$ et on peut garantir que $\lim_{\mu\to 0} \mu \frac{dH}{d\phi}(\phi,\mu)= 0$. L'analyse de stabilité est évidemment compliquée par le fait que B est critique; mais par un jeu de transformation de variable convenable et l'emploi de la méthode de Liapounoff on peut montrer que la solution périodique dont on a établi l'existence pour (13) est asymptotiquement stable si la matrice

$$S = \frac{1}{2\pi} \int_0^{2\pi} G(\phi)d\phi$$

avec $G(\phi) = R(\phi) - H(\phi)\, A^{-1}(\phi)\, Q(\phi)$ a ses valeurs propres de partie réelle négative; il y a instabilité si l'une au moins d'entre elles est de partie réelle positive.

Revenant au problème mécanique décrit par (12), sous l'hypothèse $\frac{\omega_o}{\omega_1}$ non entier, l'existence d'une solution périodique est acquise, qu'on peut représenter asymptotiquement:

$$\phi' = \omega_1 - \frac{\mu ab\omega_1^4}{2\omega_o(\omega_o^2-\omega_1^2)} \sin 2\phi + \mu\varepsilon(\mu) ,$$

$$y = \frac{ar\omega_1^2}{\omega_o^2-\omega_1^2} \cos\phi + \varepsilon(\mu) ,$$

$$z = - \frac{ar\omega_1^3}{\omega_o^2-\omega_1^2} \sin\phi + \varepsilon(\mu) .$$

Cette solution est stable si $\omega_1 < \omega_o$ ou $\omega_1 > \omega_o \sqrt{1 + \frac{a\,b}{2\,f}}$, est instable dans l'intervalle $\omega_1 \in \left] \omega_o, \omega_o\sqrt{1 - \frac{a\,b}{2\,f}} \right[$. On notera que la zone d'instabilité se rétrécit si le coefficient d'amortissement f est élévé.

Instabilité d'un rotor rigide de section anisotrope tournant a vitesse constante sur une suspension elastique anisotrope

On note Ox l'axe géométrique du rotor en position de repos horizontale, O x y z un système orthonormé de référence, O x v w le système d'axes entrainé par la rotation $\vec{\omega} = \omega \vec{x}$, où ω constante est la vitesse du rotor, supposé symétrique par rapport au plan normal à son axe en son centre d'inertie G. Le mouvement perturbé est supposé tel que G demeure confondu avec O; ainsi avec $O\,x_1\,v_1\,w_1$, système lié d'axes principaux du rotor, on peut décrire le mouvement à l'aide des variables de perturbation ϕ, Ψ : $\vec{v}_1 - \vec{v} = -\phi\,\vec{x}_1$, $\vec{w}_1 - \vec{w} = -\Psi\,\vec{x}_1$.

La tenseur d'inertie du rotor relativement aux axes $O\,x_1\,v_1\,w_1$ est $\begin{pmatrix} I_1 & & 0 \\ & I_2 & \\ 0 & & I_3 \end{pmatrix}$ et k_1, k_2 sont les rigidités des suspensions élastiques dans les directions y et z pour chacun des deux paliers centrés aux points $x = \pm \ell$.

On pose

$$\delta = \frac{I_3 - I_2}{I_3 + I_2} \quad , \quad \lambda = \frac{2I_1}{I_2 + I_3} \quad , \quad \mu = \frac{k_2 - k_1}{k_1 + k_2} \quad , \quad \omega_o^2 = 2l^2 . \frac{k_1 + k_2}{I_2 + I_3}$$

et les équations du mouvement s'écrivent [10]:

$$(14) \quad \begin{cases} (1-\delta)\psi'' + (2-\lambda)\omega\phi' - (1+\delta+\lambda)\omega^2\Psi + \omega_o^2(1+\mu\cos 2\omega t)\psi + \mu\omega_o^2 \sin 2\omega t . \phi = 0 \\ (1+\delta)\phi'' - (2-\lambda)\omega\psi' - (1-\delta-\lambda)\omega^2\phi + \omega_o^2(1-\mu\cos 2\omega t)\phi + \mu\omega_o^2 \sin 2\omega t . \psi = 0 . \end{cases}$$

On suppose le rotor long c'est-a-dire $I_2 > I_1$, $I_3 > I_1$ qui implique $1 + \delta - \lambda > 1 - \delta - \lambda > 0$.

Dans le cas d'une suspension isotrope $\mu=0$, on a un système avec couplage gyroscopique classique dont la zone d'instabilité est:

$$(1+\delta-\lambda)^{-1} < \frac{\omega^2}{\omega_o^2} < (1-\delta-\lambda)^{-1} .$$

Mais, parce que le frottement, qoique très réduit, n'est jamais totalement absent, on peut prévoir d'après la théorie classique que le système sera surement instable dans la zone $(1+\delta-\lambda)^{-1} < \frac{\omega^2}{\omega_o^2}$, d'où l'existence d'une valeur critique supérieure pour la rotation ω. Mais si $\mu \neq 0$, des zones additionnelles d'instabilité peuvent apparaître en régime subcritique, en raison du phénomène d'excitation paramétrique. La référence déjà citée propose une approche numérique du problème; mais l'on peut aussi y apporter une réponse analytique sur la base de certaines idées avancées plus haut, comme nous allons le montrer maintenant.

Il est approprié de récrire (14) en terme de nouvelles variables $\psi \to \sqrt{1+\delta}\,\Psi$, $\phi \to \sqrt{1-\delta}\,\Phi$, c'est-à-dire:

(15)
$$\Psi'' + \frac{2-\lambda}{\sqrt{1-\delta^2}}\,\omega\,\Phi' + (\omega_o^2 - (1+\delta-\lambda)\omega^2)\,\frac{\Psi}{1-\delta} + \mu\omega_o^2\left[\frac{\Psi}{1-\delta}\cos 2\omega t + \frac{\Phi}{\sqrt{1-\delta^2}}\,.\sin 2\omega t\right] = 0$$

$$\Phi'' - \frac{2-\lambda}{\sqrt{1-\delta^2}}\,\omega\Psi' + (\omega_o^2 - (1-\delta-\lambda)\omega^2)\,\frac{\Phi}{1+\delta} + \mu\omega_o^2\left[\frac{\Psi}{\sqrt{1-\delta^2}}\sin 2\omega t - \frac{\Phi}{1+\delta}\,.\cos 2\omega t\right] = 0\,.$$

Le système (15) est réciproque, c'est-à-dire que si $\Phi(t), \Psi(t)$ en est solution, alors $-\Phi(-t)$, $\Psi(-t)$ est aussi une solution; il en résulte que, pour chaque ω, il y aura stabilité de la solution nulle pourvu que μ soit assez petit, à condition que:

(16)
$$2\omega_1 \neq 0 \quad , \quad 2\omega_2 \neq 0 \quad , \quad \omega_1 \pm \omega_2 \neq 0 \qquad \text{mod } 2\omega$$

où ω_1, ω_2 sont les pulsations propres, d'ailleurs dépendant de ω, du système gyroscopique décrit par (15) quand $\mu = 0$. Les valeurs subcritiques de ω, pour lesquelles $2\,\omega_1 = 0$, ou $2\,\omega_2 = 0$ ou $\omega_1 \pm \omega_2 = 0$ mod $2\,\omega$, définissent sur l'axe ω les points d'arrêt des zones d'instabilité que l'on désire calculer dans le plan (μ,ω).

On peut faire une étude complète mais on se bornera dans ce qui suit à présenter les résultats du calcul pour la résonance $\omega = \frac{\omega_1+\omega_2}{2}$.

Avec $\sigma = \dfrac{2-\lambda}{2\sqrt{1-\delta^2}} > 0$ et posant

(17) $$y = \left(\frac{\omega_o}{\omega}\right)^2 + \lambda - 2$$

l'équation qui définit les valeurs admissibles de y pour cette résonance est:

(18) $$\delta^2 y^2 + 4(\sigma^2-1)(1-\delta^2)y + 4\,\sigma^2(\sigma^2-1)(1-\delta^2)^2 = 0 .$$

Si $\sigma^2 < \dfrac{2+\delta}{2(1+\delta)}$ (cas du rotor long, qui sera retenu ici),

seule la racine négative de (18) convient, conduisant par (17) à une valeur subcritique $\tilde{\omega}$ de la rotation, tandis que la racine positive serait associée à la résonance $2\,\omega = \omega_2 - \omega_1$.

Pour étudier la stabilité au voisinage du point $\mu = 0$, $\omega = \tilde{\omega}$, il est approprié d'introduire une représentation $\omega = \tilde{\omega}\,(1+\mu\,\gamma)^{-\frac{1}{2}}$ et de prendre désormais pour variables:

$$\tau = 2\omega t \quad , \quad \dot{\psi} = \frac{d\Psi}{d\tau} \quad , \quad \dot{\phi} = \frac{d\phi}{d\tau}$$

et substituer à μ le paramètre $\upsilon = \dfrac{\mu}{4}\left(\dfrac{\omega_o}{\tilde{\omega}}\right)^2$.

Les équations (15) peuvent alors être récrites sous la forme:

(19) $$\begin{aligned} \ddot{\Psi} + \sigma\dot{\phi} + p\Psi &= \upsilon f \\ \ddot{\phi} - \sigma\dot{\psi} + q\phi &= \upsilon g \end{aligned}$$

avec

$$f = -\,\frac{\gamma+\cos\tau}{1-\delta}\,\Psi - \frac{\sin\tau}{\sqrt{1-\delta^2}}\,\phi + O(\upsilon) \,,$$

$$g = \frac{\cos\tau-\gamma}{1-\delta}\,\Phi - \frac{\sin\tau}{\sqrt{1-\delta^2}}\,\Psi + O(\upsilon)$$

et

$$4(1-\delta)p = \left(\frac{\omega_o}{\tilde{\omega}}\right)^2 - (1+\delta-\lambda)$$

(20)

$$4(1+\delta)q = \left(\frac{\omega_o}{\tilde{\omega}}\right)^2 - (1-\delta-\lambda)\,.$$

Quand $\upsilon = 0$ les pulsations propres de (19) sont w_1, w_2 racines de $w^4 - (\sigma^2+p+q)\,w^2 + pq = 0$, telles que $w_1 + w_2 = 1$ (condition de résonance).

On utilise une méthode de modulation d'amplitude à partir d'une représentation

$$\Psi = \xi_1 \sin w_1\tau + \eta_1 \cos w_1\tau + \xi_2 \sin w_2\tau + \eta_2 \cos w_2\tau$$

(21)

$$\Phi = k(\xi_1 \cos w_1\tau - \eta_1 \sin w_1\tau) + r\,(\xi_2 \cos w_2\tau - \eta_2 \sin w_2\tau)$$

en termes de variables lentes ξ_i, η_i $i = 1,2$, avec

$$k = \frac{\sigma w_1}{q-w_1^2} = \frac{p-w_1^2}{\sigma\, w_1} \quad , \quad r = \frac{p-w_2^2}{\sigma\, w_2} = \frac{\sigma w_2}{q-w_2^2}$$

et imposant que

$$\dot{\Psi} = \xi_1 w_1 \cos w_1\tau - \eta_1 w_1 \sin w_1\tau + \xi_2\, w_2 \cos w_2\tau - \eta_2 w_2 \sin w_2\tau$$

(22)

$$\dot{\Phi} = -kw_1(\xi_1 \sin w_1\tau + \eta_1 \cos w_1\tau) - r\, w_2(\xi_2 \sin w_2\tau + \eta_2 \cos w_2\tau)\,.$$

Introduisant par ailleurs les variables $\Theta_1 = w_1\tau$, $\Theta_2 = w_2\tau$, $\tau = \Theta_1 + \Theta_2$, il advient de (19), (21), (22):

$$\dot{\xi}_1 = \upsilon\left[\frac{r\cos\Theta_1}{rw_1-kw_2}\ f - \frac{\sin\Theta_1}{kw_1-rw_2}\ g\right],$$

$$\dot{\xi}_2 = \upsilon\left[-\frac{k\cos\Theta_2}{rw_1-kw_2}\ f + \frac{\sin\Theta_2}{kw_1-rw_2}\ g\right], \qquad \frac{d\Theta_1}{dt} = w_1,$$

$$\dot{\eta}_1 = \upsilon\left[-\frac{r\sin\Theta_1}{rw_1-kw_2}\ f - \frac{\cos\Theta_1}{kw_1-rw_2}\ g\right], \qquad \frac{d\Theta_2}{dt} = w_2,$$

$$\dot{\eta}_2 = \upsilon\left[\frac{k\sin\Theta_2}{rw_1-kw_2}\ f + \frac{\cos\Theta_2}{kw_1-rw_2}\ g\right],$$

$$f = -\Bigg\{\left(\frac{\gamma+\cos(\Theta_1+\Theta_2)}{1-\delta}\sin\Theta_1 + \frac{k\sin(\Theta_1+\Theta_2)}{\sqrt{1-\delta^2}}\cos\Theta_1\right)\xi_1 + \left(\frac{\gamma+\cos(\Theta_1+\Theta_2)}{1-\delta}\cos\Theta_1\right.$$

$$\left. - \frac{k\sin(\Theta_1+\Theta_2)}{\sqrt{1-\delta^2}}\sin\Theta_1\right)\eta_1 + \left(\frac{(\gamma+\cos(\Theta_1+\Theta_2)}{1-\delta}\sin\Theta_2 + \frac{r\sin(\Theta_1+\Theta_2)}{\sqrt{1-\delta^2}}\cos\Theta_2\right)\xi_2$$

$$+\left(\frac{(\gamma+\cos(\Theta_1+\Theta_2)}{1-\delta}\cos\Theta_2 - \frac{r\sin(\Theta_1+\Theta_2)}{\sqrt{1-\delta^2}}\sin\Theta_2\right)\eta_2\Bigg\} + O(\upsilon),$$

$$g = -\Bigg\{\left(k\frac{\gamma-\cos(\Theta_1+\Theta_2)}{1+\delta}\cos\Theta_1 + \frac{\sin(\Theta_1+\Theta_2)}{\sqrt{1-\delta^2}}\sin\Theta_1\right)\xi_1 + \left(-k\frac{\gamma-\cos(\Theta_1+\Theta_2)}{1+\delta}\sin\Theta_1\right.$$

$$\left. + \frac{\sin(\Theta_1+\Theta_2)}{\sqrt{1-\delta^2}}\cos\Theta_1\right)\eta_1 + \left(r\frac{\gamma-\cos(\Theta_1+\Theta_2)}{1+\delta}\cos\Theta_2 + \frac{\sin(\Theta_1+\Theta_2)}{\sqrt{1-\delta^2}}\sin\Theta_2\right)\xi_2$$

$$+ \left(- r \frac{\gamma - \cos(\Theta_1 + \Theta_2)}{1+\delta} \sin\Theta_2 + \frac{\sin(\Theta_1+\Theta_2)}{\sqrt{1-\delta^2}} \cos\Theta_2\right)\eta_2 \Bigg\} + O(\upsilon),$$

Prenant les moyennes des seconds membres par rapport aux variables indépendantes Θ_1, Θ_2, sous l'hypothèse que w_1, w_2 sont des pulsations indépendantes, on obtient le système:

$$\dot{\xi}_1 = \upsilon(-\gamma\alpha\eta_1 + a\eta_2)$$

$$\dot{\xi}_2 = \upsilon(b\eta_1 + \gamma\beta\ \eta_2)$$

$$\dot{\eta}_1 = \upsilon(\gamma\alpha\xi_1 + a\xi_2)$$

$$\dot{\eta}_2 = \upsilon(b\xi_1 - \gamma\beta\xi_2)$$

à propos duquel on détermine facilement la condition d'instabilité de la solution nulle

$$\gamma^2 < \frac{4ab}{(\alpha-\beta)^2} = \gamma^{*2}$$

et on trouve après calcul:

$$\gamma^* = \frac{\left(\sigma \frac{1+\delta}{1-\delta} - \left(\sqrt{\frac{p}{q}} - 1\right)\sqrt{\frac{1+\delta}{1-\delta}} - \sigma\sqrt{\frac{p}{q}}\right)\left(\frac{p}{q}\right)^{1/4}}{\left(\sqrt{\frac{p}{q}} + 1\right)\left(1-4\sqrt{pq}\right)^{1/2}\left(\frac{1+\delta}{1-\delta} + \sqrt{\frac{p}{q}}\right)}$$

avec p,q définis par (20), $\omega = \tilde{\omega}$ étant par (17) associé à la racine négative de (18).

Dès lors on peut dire que pour $\mu > 0$ assez petit, il y aura instabilité de la solution nulle de (15) pour $|\gamma| < \gamma^*$ avec $\omega = \tilde{\omega}\,(1+\mu\gamma)^{-\frac{1}{2}}$; la largeur de la zone d'instabilité est dans ce cas $O(\mu)$.

Ajoutons pour conclure que la résonance $\omega = \omega_1$ $(\omega_1 < \omega_2)$ est atteinte en régime subcritique pour une vitesse ω moindre que celle qui correspond à la résonance $2\omega = \omega_1 + \omega_2$; le calcul montre aussi que la largeur de la zone d'instabilité est $O(\mu^2)$.

References

[1] C.L. SIEGEL et J.K. MOSER: Lectures on Celestial Mechanics. Springer 1971.

[2] BOGOLIOUBOFF, MITROPOLSKY, SAMOILENKO: Methods of accelerated convergence in non linear mechanics. Springer 1976.

[3] J.K. MOSER: Convergent series expansions for quasiperiodic motions. Math. Annalen, vol. 169, 1967, pp.136-176.

[4] M. ROSEAU: Régimes quasi périodiques dans les systèmes vibrants non linéaires. J. Math. pures et appl., vol. 57, 1978, pp.21-68.

[5] M. ROSEAU: La méthode de modulation d'amplitude et son application à l'étude des oscillateurs couplés. Journal de Mécanique, vol. 20, 1981, pp.199-217.

[6] M. ROSEAU: On the coupling between a vibrating mechanical system and the external forces acting upon it. Int. Journal of non linear mechanics, 1982.

[7] Y. ROCARD: Dynamique générale des vibrations. Masson 3ème ed., Paris 1960.

[8] G. PANOVKO, I.I. GUBANOVA: Stability and oscillations of elastic systems. N.Y. consultant bureau, 1965.

[9] M. ROSEAU: Some cases of instability in rotating machinery; an approach based on the theory of singular perturbation. IX International Conference on nonlinear oscillations, Kiev U.R.S.S., 1981.

[10] G.M.L. GLADWELL, C.W. STAMMERS: Prediction of instable regions of a reciprocal system governed by a set of linear equations. J. Sound Vibrations, vol. 8, 1968, pp.457-468.